THEORY OF DISCRETE AND CONTINUOUS FOURIER ANALYSIS

H. Joseph Weaver

WILEY

A WILEY-INTERSCIENCE PUBLICATION

JOHN WILEY & SONS

New York • Chichester • Brisbane • Toronto • Singapore

Library of Congress Cataloging in Publication Data:

Weaver, H. Joseph.
    Theory of discrete and continuous Fourier analysis/H. Joseph
Weaver.
        p. cm.
    "A Wiley-Interscience publication."
    Includes bibliographies and index.
    ISBN 0-471-62872-7
    1. Fourier analysis.   I. Title.
QA403.5.W44 1988
515'.2433—dc19   88-10820
                                                                CIP

Printed in the United States of America

10 9 8 7 6 5 4 3 2

PREFACE

The expression *Discrete and Continuous Fourier Analysis* as used in this work means the combination of three distinct but notably similar fields. These three fields are the Fourier series, the Fourier transform and the discrete Fourier transform. These subjects are rarely (if ever) combined and offered together as a single course. More often than not, the student learns about them as a subtopic in a broader course. For example, the Fourier series is usually first introduced as a section in an intermediate calculus course. One of the main purposes of this text is to consolidate the presentation of these subjects. When this is accomplished the student should be readily able to appreciate their analogous properties and hopefully gain additional insight into one topic from a knowledge of the others.

In an attempt to make this text self-contained I have included two chapters on basic mathematical analysis and integration theory along with a third one dealing with an introduction to the theory of distributions. Readers who are already familiar with these topics may wish to skip these chapters and proceed directly to the remaining ones which deal specifically with Fourier analysis.

This is my second work published by Wiley dealing with the subject of Fourier analysis. The first book titled *Applications of Discrete and Continuous Fourier Analysis* was designed as a reference book to be used mainly in applying Fourier analysis to a wide assortment of scientific and engineering fields. The approach taken in that first work was to present just enough theory to allow the reader to understand the basic concepts in order to be able to perform meaningful and efficient manipulations with Fourier analysis. This text presents a more theoretical study of Fourier analysis and is meant to serve as both a reference and a text book. As a reference it complements the first text by providing the reader with a deeper understanding of the properties and behavior of the Fourier transform and the Fourier series. As a text it provides a reasonably thorough and understandable treatment of the theory of Fourier analysis. There is sufficient material in this text for a one semester course on the subject.

Even though this text is advertised as a theoretical treatment of the subject it is purposely filled with a large number of illustrative example problems to help elucidate the applied side of the theory. In addition, every chapter contains a set of homework problems that can be worked by the readers to test their understanding of the material presented. In addition these problem sets were designed to supplement the material presented in the chapter.

The chapter contents of this text are briefly summarized as follows:

Chapter 1 Introduction. This chapter serves as an introduction to both the Fourier series and the Fourier transform. Several examples are presented that point out the usefulness of (and the motivation for) a theoretical study of Fourier analysis.

Chapter 2 Basic Mathematical Background. This chapter presents a review of several basic concepts from mathematical analysis. It includes such topics as: functions, sequences, metric spaces, limits, continuity, convergence, and so on. These are subjects that are very helpful in the understanding of Fourier analysis. This chapter is included for the sake of completeness and may be omitted by the reader who is already familiar with the material.

Chapter 3 Integration Theory. The concept of integration plays a very important role in the theory of the Fourier series and the Fourier transform. Both require the evaluation of integral expressions which are to be considered in the Lebesgue sense. This chapter introduces such topics as step functions, Lebesgue measure, integrable functions, square-summable function, series of integrable functions, and so on. Again this chapter is included for completeness and may be skipped by readers already knowledgeable in this area.

Chapter 4 Distribution Theory. Many of the more useful functions, such as the delta and comb, cannot be adequately described using standard mathematical analysis. In this chapter the inner product of two functions is used as a guide to introduce the concept of a distribution, which provides a convenient and compact method of handling the nonconformist functions. A distribution is then defined as a mapping from a particular set of functions to the set of complex numbers.

Chapter 5 The Fourier Series. This chapter begins the formal treatment of Fourier analysis and discusses the Fourier series of periodic functions that are assumed to be *square-summable* and/or *Lebesgue integrable*. The main focus in this chapter is on the Fourier series representation of a function in terms of the trigonometric system of orthogonal functions. However, the chapter begins with a broader discussion of a general system of orthogonal functions and their properties. The Dirichlet kernel function is introduced and used to discuss convergence of the Fourier series as well as Gibbs' phenomenon. Several very useful theorems are presented that can often be used to facilitate the actual calculation of the series coefficients. Numerous examples are presented throughout this chapter. These examples are chosen in such a way as to illustrate many of the theoretical concepts presented in the chapter.

Chapter 6 The Fourier Transform. This chapter introduces the Fourier transform of a class of functions known as *absolutely integrable functions*. It begins with a discussion of existence and proceeds to consider some of the unique behavior and characteristics of the transform of such functions. Several theorems are presented which are particularly useful for calculational and manipulational purposes. Convolution and correlation of functions are discussed as well as the powerful Convolution Theorem. As in the previous chapter, a large number of solved example problems are used to illustrate the theoretical concepts.

Chapter 7 Fourier Transform of a Distribution. This chapter presents the Fourier transform in terms of distribution theory. This approach permits the calculation of the transform of a much larger class of functions (distributions) than those presented in the previous chapter. It is shown that most of the useful property theorems previously demonstrated for absolutely integrable functions are also valid for just about any plausible function. Furthermore, it is shown that when convolution is revisited in the light of distribution theory new insight into the subject can be obtained. Also, this chapter establishes a simple and clear relationship between the Fourier transform and the Fourier series. Again, theoretical concepts are clarified with illustrative example problems.

Chapter 8 Discrete Fourier Transform. This chapter presents the discrete Fourier transform which is a mapping between two Nth order sequences. The striking similarities between the Fourier transform and the discrete Fourier transform are explored via several property theorems. Common misconceptions concerning periodicity requirements and the discrete Fourier transform are resolved in this chapter. Convolution of sequences and the Convolution theorem for sequences are presented. Finally this chapter ends with a brief description of the fast Fourier transform algorithm.

Chapter 9 Sampling Theory. Probably the most useful aspect of the discrete Fourier transform is the fact that it can be used to accurately approximate the Fourier transform (and Fourier series) of an arbitrary function or waveform. This is accomplished by converting the function to a sequence (via sampling) and then recovering the Fourier transform of the function from the discrete Fourier transform of the sampled sequence. In this chapter sampling is presented as the product of the original function and a comb function. When this point of view is examined in the frequency domain much insight into sampling theory, as well as a clean proof of the celebrated Whittaker-Shannon sampling theorem, can be obtained. Problems with aliasing and windowing are addressed in both the time and frequency domain. The theoretical discussions offered in this chapter are summarized as a six step procedure that can be used to obtain the Fourier transform of a function by employing sampling theory and the discrete Fourier transform. This same six step procedure has been implemented as a FORTRAN computer subroutine and is presented in the Appendix.

A portion of the material presented in this text was developed during the 1970's for a class that I presented at Lawrence Livermore National Laboratory as part of their Continuing Education Program. I happily acknowledge the encouragement of Mr. Al Cassell who was very involved and helpful in this effort. I am also grateful to several colleagues at Livermore who took the time to review and comment on the present manuscript. In particular, special thanks are due to Mr. R. Bradley Burdick, and Mr. Thomas A. Biesiada for their efforts. The development of the treatment on Gibbs' phenomena was strongly influenced by comments received from Dr. Richard A. Forman of the National Bureau of Standards. Additionally, Dr. Richard W. Hamming of the Naval Postgraduate School provided valuable insights and comments on the same subject. Finally, I would like to thank Dr. Tom Mincer of the Continuing Education Institute for his support and sponsorship of my short courses on Fourier analysis during the last several years.

I am most grateful to my wife Sue for her patience and understanding during the past few years while I completed this manuscript for publication.

H. JOSEPH WEAVER

Livermore, California
September 1988

To My Parents
Harold and Winifred

CONTENTS

1 Introduction   1

The Fourier Series   3
The Fourier Transform   11
The Impulse Function   17
Summary   19
Problems   19
Bibliography   20

2 Basic Mathematical Background   21

Fundamental Set Theory   21
Set Algebra   22
Laws of Internal Composition   25
Special Elements of a Set   27
Groups, Fields, and Vector Spaces   28
Functions, Relations, and Sequences   32
Distance and Metric Spaces   35
Concept of Order   37
Concepts from Topology   39
Limits and Continuity   40
Convergence of Sequences   43
Choice of a Norm for Convergence   44
Cauchy Convergence   47
Summary   48
Problems   48
Bibliography   50

3 Integration Theory   51

Lebesgue Measure   51
Step Functions and Their Integrals   54
Lebesgue Functions   57
Riemann and Lebesgue Integrals   64
Convergence of Sequences of Lebesgue Integral Functions   65
Lebesgue Functions on an Unbounded Interval   69
Integral Functions   72
Square-Summable Functions   76
Summary   78
Problems   79
Bibliography   80

4 Distribution Theory   81

Inner Product of Two Functions   81
Existence of the Inner Product   83
Definition of a Distribution   85
Properties of Distributions   87
The Comb Distribution   89
Derivative of a Distribution   91
Odd and Even Distributions   94
The Delta Distribution: An Intuitive Approach   94
Summary   97
Problems   98
Bibliography   99

5 The Fourier Series   101

General System of Orthogonal Functions   103
Minimum Mean Square Error   104
Bessel's Inequality and Convergence in the Mean   106
Properties of Complete Systems   108
Trigonometric System of Orthogonal Functions   110
Complex Exponential System of Orthogonal Functions   115
The Riemann Lebesgue Lemma   116
The Dirichlet Kernel Function   117
Pointwise Convergence of the Fourier Series   121
Gibbs' Phenomenon   125
Fourier Series Representation of Complex Functions   130
Properties of the Fourier Series   131
Examples   138
Summary   149
Problems   149
Bibliography   151

6 The Fourier Transform   **153**

Existence of the Fourier Transform   154
Properties and Behavior of the Fourier Transform   159
The Shifting Theorems   163
The Derivative Theorems   170
Transform of a Transform   173
Symmetry Considerations   174
Uniqueness and Reciprocity   178
Convolution of Two Functions   181
Correlation   193
Self-Reciprocity and the Hermite Functions   198
Summary   202
Problems   203
Bibliography   205

7 Fourier Transform of a Distribution   **207**

Linearity and Scale Change   210
The Shifting Theorems   211
The Derivative Theorems   214
Symmetry Considerations   219
Fourier Transform of the Comb Distribution   221
Convolution of Distributions   224
Physical Interpretation of Convolution   227
Fourier Transform of a Periodic Function: The Fourier Series   230
Summary   232
Problems   233
Bibliography   234

8 The Discrete Fourier Transform   **235**

Nth-Order Sequences   235
The Discrete Fourier Transform   237
Properties of the Discrete Fourier Transform   243
Symmetry Relations   253
Convolution of Two Sequences   257
Simultaneous Calculation of Real Transforms   260
The Fast Fourier Transform   263
Summary   266
Problems   266
Bibliography   267

9 Sampling Theory   **269**

Sampling a Function   270
Aliasing   273

Computer Calculation of the Fourier Transform   274
Computer-Generated Fourier Series   280
Super-Gaussian Windows   282
Summary   293
Problems   294
Bibliography   295

Appendix: Fourier Transform FORTRAN Subroutine   297

Index   303

1

INTRODUCTION

It's hard to think of any other branch of mathematics that has as many practical applications as Fourier analysis. The mathematics of Fourier analysis (the Fourier series and Fourier transform) has been used for many years to study physical phenomena from a wide variety of scientific and engineering fields. Some examples that come readily to mind are heat transfer, wave propagation, circuit analysis, vibrations, control systems analysis, optical systems, and electronic circuit analysis. It is interesting to note that one reason Fourier analysis has such a wide span of applications is that the Fourier kernel, $\exp[2\pi iwt]$, is a solution to an nth-order linear differential equation which, in turn, is used to model the above-mentioned physical phenomena.

In addition to being a very useful and practical branch of mathematics, Fourier analysis has also been responsible for the development of much of the modern theory of mathematics. This was brought about mainly because a large portion of the early work was based upon the intuition and cleverness of its users. For example, Bernoulli (when studying vibrational motion of a string) reasoned that an infinite linear combination of sine functions could be combined to represent a continuous function over the interval $[0, \pi]$. Pure mathematicians of the time scoffed at this and were unwilling to accept this conclusion. However, when the mathematical predictions began to match physical reality, mathematical theory had to be modified and improved in order to explain the agreement.

Actually, many of the early "mechanical" manipulations of Fourier analysis pushed the theoretical development of the subject forward. Application-minded users were interested in using the mathematics to solve real-world problems. They were not particularly concerned with such theoretical issues as why or when a Fourier series converged to the generating function. Even though they used integration techniques to develop their answers, they did not dwell upon existence criteria for the integrals that they were calculating. This was to be the task of the pure mathematicians.

As work progressed in the field and people began to find other applications, various useful (but ill-behaved) functions such as the delta and comb functions were introduced. These were particularly appropriate in describing such things as point charges, point loads, impulse excitations, and so on. These functions were essentially impossible to describe using standard mathematical function theory, and, consequently, *distribution theory* was developed.

Throughout its 200-plus year history, Fourier analysis has been steadily advanced. Before the advent of digital computers, Fourier transforms (and series) were calculated analytically. A digital computer, coupled with the discrete Fourier transform (DFT), made it theoretically possible to calculate both the Fourier transform and Fourier series of any "reasonably well-behaved" function. Unfortunately, because a very large number of mathematical operations (complex additions and multiplications) were required by the DFT, it was not really practical. In the mid-1960s the "fast Fourier transform" (FFT) was conceived and things began to change. The FFT is not really a transform but, instead, an efficient algorithm for calculating the DFT of a sequence. The FFT dramatically reduced the number of mathematical operations required to calculate the discrete Fourier transform of a sequence. Also, by the mid-1970s, smaller, faster, and less expensive minicomputers were being produced. This fact, coupled with the highly efficient FFT algorithm, led to a renewed interest in using Fourier analysis. By using digital computers to calculate, or more precisely approximate, the Fourier transform of a function, one was no longer limited to studying functions that possessed an analytical expression which could be integrated. As a matter of fact, when a digital computer is used, the function doesn't even have to be described by a mathematical expression. Such would be the case when studying the frequency behavior of functions generated by electronic equipment monitoring a specific physical experiment. For example, such signals are obtained routinely when studying vibrational behavior of structures using accelerometers.

One might conclude that since digital computers allow us to obtain the Fourier transform of just about any physically realizable function, there is actually no longer any need to consider the analytical behavior, or theoretical aspects, of Fourier analysis. Actually, just the opposite is true. Inasmuch as a digital computer will almost always produce an "answer" when programmed to transform a sampled function we must have a good understanding of Fourier analysis to judge if the "answer" is correct or even reasonable. In other words, the FFT can be used as a "black box" to produce Fourier transforms and Fourier series. Therefore, to avoid the risk of misapplication of the results, the user should have some reasonable knowledge of the basic theory and behavior of the Fourier transformations.

In this text we take a look at some of the mathematical theory that is associated with the Fourier series and Fourier transform. This presentation will be given with the stipulated purpose of helping the user of Fourier analysis be more aware of what limitations must be considered when performing various calculations and manipulations. Our presentation will be limited to *trigonometric* Fourier analysis. Specifically, we will deal with the Fourier series, Fourier transform, and discrete Fourier transform. All three may be considered Fourier mappings. The remainder of this chapter is devoted to providing the motivation for considering the more theoretical aspects of Fourier analysis.

THE FOURIER SERIES

The Fourier series of a periodic function (with period T) is defined, in rectangular form, as

$$f(t) = \sum_{k=0}^{\infty} \left[A_k \cos\left(\frac{2\pi k t}{T}\right) + B_k \sin\left(\frac{2\pi k t}{T}\right) \right], \qquad (1.1)$$

where the Fourier series coefficients A_k and B_k are given by

$$A_k = \frac{2}{T} \int_{-T/2}^{T/2} f(t) \cos\left(\frac{2\pi k t}{T}\right) dt, \qquad k = 1, 2, \ldots, \infty,$$

$$A_0 = \frac{1}{T} \int_{-T/2}^{T/2} f(t) \, dt, \qquad (1.2)$$

$$B_k = \frac{2}{T} \int_{-T/2}^{T/2} f(t) \sin\left(\frac{2\pi k t}{T}\right) dt, \qquad k = 1, 2, \ldots, \infty,$$

$$B_0 = 0. \qquad (1.3)$$

The Fourier series representation of a function can also be placed in a complex exponential form that often simplifies the mathematical manipulations. This form also more closely resembles the Fourier transform representation of a function. This is discussed in greater detail in Chapter 5.

Let us now perform a few example calculations of the Fourier series of different functions to illustrate the mathematical manipulations as well as some of the problems and/or dilemmas that we can arrive at.

The first function that we deal with is the *triangle sawtooth* function illustrated in Figure 1.1 and described mathematically over the domain $[-1, 1]$ as:

$$f(t) = \begin{cases} 1 + t, & t \in [-1, 0), \\ 1, & t = 0, \\ 1 - t, & t \in (0, 1]. \end{cases}$$

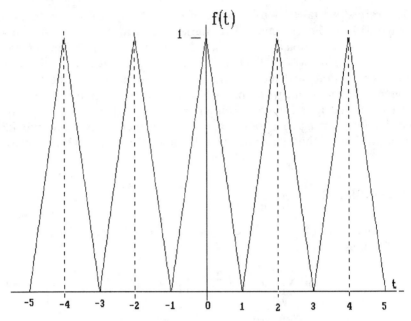

Figure 1.1 The sawtooth function.

As we can see from Figure 1.1, we have made the assumption that our period extends from -1 to $+1$. Obviously, for this choice of an interval we have the period width $T = 2$. First using equation (1.2) we find

$$A_k = \frac{2}{T} \int_{-T/2}^{T/2} f(t)\cos\left(\frac{2\pi kt}{T}\right) dt, \qquad k = 1,2,\ldots,\infty,$$

$$A_k = \frac{2}{2} \int_{-1}^{1} f(t)\cos\left(\frac{2\pi kt}{2}\right) dt, \qquad k = 1,2,\ldots,\infty,$$

$$A_k = \int_{-1}^{0} (1+t)\cos(\pi kt)dt + \int_{0}^{1} (1-t)\cos(\pi kt)dt, \qquad k = 1,2,\ldots,\infty,$$

$$A_k = \frac{2 - 2\cos(\pi k)}{(\pi k)^2}, \qquad k = 1,2,\ldots,\infty.$$

Also,

$$A_0 = \frac{1}{T} \int_{-T/2}^{T/2} f(t)dt = \frac{1}{2} \int_{-1}^{0} (1+t)dt + \frac{1}{2} \int_{0}^{1} (1-t)dt = \frac{1}{2}.$$

As for the sine term coefficients B_k, we use equation (1.3) to obtain

$$B_k = \frac{2}{T} \int_{-T/2}^{T/2} f(t) \sin\left(\frac{2\pi k t}{T}\right) dt, \qquad k = 1, 2, \ldots, \infty,$$

$$B_k = \int_{-1}^{0} (1 + t) \sin(\pi k t) dt + \int_{0}^{1} (1 - t) \sin(\pi k t) dt = 0.$$

When the above expressions for A_k, A_0, and B_k are substituted into equation (1.1) we find

$$f(t) = \frac{1}{2} + \sum_{k=1}^{\infty} \left[\frac{2 - 2\cos(\pi k)}{(\pi k)^2}\right] \cos(\pi k t).$$

We note that the term in brackets only has a value when k is odd and assumes the value $4/(\pi k)^2$. When k is even, $A_k = 0$. Thus, the above expression could be somewhat simplified. However, at this point we choose to leave it in this more general form.

The summation of equation (1.1) is an infinite summation that ranges from 0 to ∞. While such an infinite summation is of theoretical interest, it is obviously not possible to physically realize or calculate. For this reason we also use the *Fourier series approximation* of a function, which is defined as the following summation:

$$f(t) \sim \sum_{k=0}^{N} A_k \cos\left(\frac{2\pi k t}{T}\right) + B_k \sin\left(\frac{2\pi k t}{T}\right). \tag{1.4}$$

Note that in equation (1.4) we are using a finite number of terms of the Fourier series to approximate the function. This is where convergence issues normally come into play. Even in the theoretical case, when we use an infinite number of terms, the series may not converge exactly to the generating function $f(t)$. Again, this is discussed in more detail in Chapter 5.

Shown in Figure 1.2a–c are the Fourier series approximations to the sawtooth function for $N = 3$, $N = 9$, and $N = 30$, respectively. As we can see from this figure, the more terms of the series that are used in the approximation, the better the function seems to converge to the generating sawtooth function. It is also worthwhile to note that the sawtooth function that we used in this example is a rather well-behaved and continuous function.

Next let us consider the boxcar function which contains jump discontinuities. The *boxcar* function is described mathematically as

$$f(t) = \begin{cases} 1, & t \in [-a, a] \\ 0, & \text{otherwise.} \end{cases}$$

For this function we assume that the period extends from -1 to $+1$. For calculational purposes we will also choose a particular value of $a = \frac{1}{2}$. This function is shown graphically in Figure 1.3.

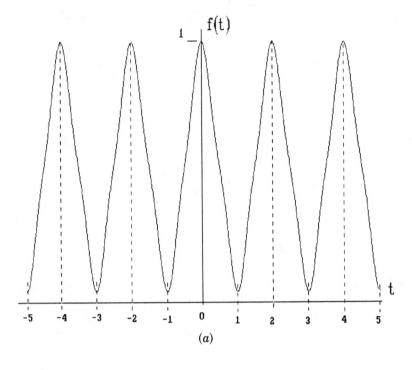

(a)

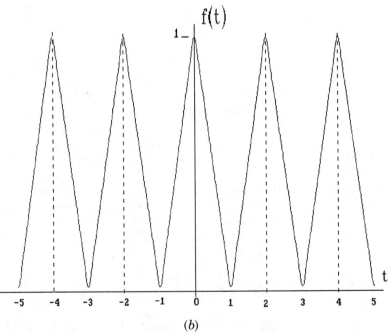

(b)

Figure 1.2 Fourier series approximation to sawtooth function. (a) Three terms, (b) nine terms, (c) 30 terms.

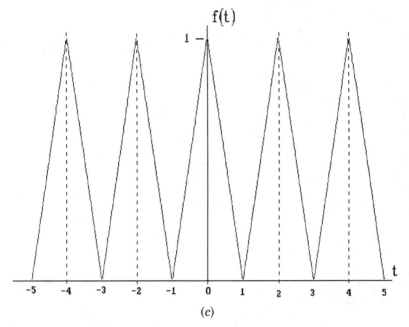

Figure 1.2   *(Continued)*

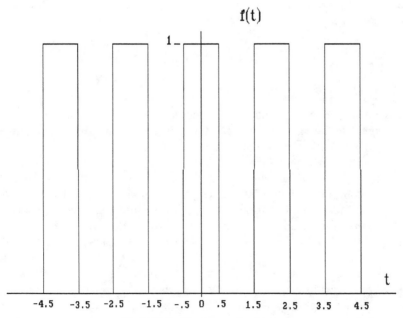

Figure 1.3   The boxcar function.

Again, the calculation of the Fourier series coefficients is a straightforward application of equations (1.2) and (1.3), that is,

$$A_k = \frac{2}{2}\int_{-1}^{1} f(t)\cos\left(\frac{2\pi k t}{2}\right) dt, \qquad k = 1,2,...,\infty,$$

$$A_k = \int_{-1/2}^{1/2} \cos(\pi k t)\,dt = \left.\frac{\sin(\pi k t)}{\pi k}\right|_{-1/2}^{1/2},$$

$$A_k = \frac{2\sin(\pi k/2)}{\pi k} = \frac{\sin(\pi k/2)}{\pi k/2} = \mathrm{sinc}\left(\frac{\pi k}{2}\right).$$

The A_0 or "DC" term is calculated to be

$$A_0 = \tfrac{1}{2}\int_{-1}^{1} f(t)\,dt = \tfrac{1}{2}\int_{-1/2}^{1/2} dt = \tfrac{1}{2}.$$

Application of equation (1.3) reveals that all of the sine term coefficients B_k are zero, that is,

$$B_k = 0 \qquad \text{for all } k.$$

Substitution of these expressions for A_k, A_0, and B_k into equation (1.4) for the Fourier series approximation, we find that the boxcar function is represented by

$$f(t) \sim \frac{1}{2} + \sum_{k=1}^{N} \mathrm{sinc}\left(\frac{\pi k}{2}\right)\cos(\pi k t).$$

Shown in Figure 1.4a–c are the Fourier series approximations to this function with $N = 3$, $N = 9$, and $N = 30$ terms, respectively. From this figure we can appreciate that as the number of terms increases, the series approximation seems to approach the boxcar generating function. However, we also note that something is indeed "happening" at the edges, or discontinuities, of the function. At this location the approximation tends to oscillate or ring. This phenomenon is known as *Gibbs' effect* and is discussed in Chapter 5, where an in-depth description of the Fourier series is presented. Here, however, we heuristically argue that this is caused because we are attempting to approximate a discontinuous function with a series of continuous (sinusoidal) functions.

This example illustrates some interesting points. First, we note that the evaluation of the integral expressions to obtain the Fourier series coefficients A_k was very straightforward. In other words, they existed and were, in fact, trivial to calculate. However, even though the integral expressions existed, it appears that something is going wrong when we form a linear combination of the Fourier series components to attempt to approximate the generating

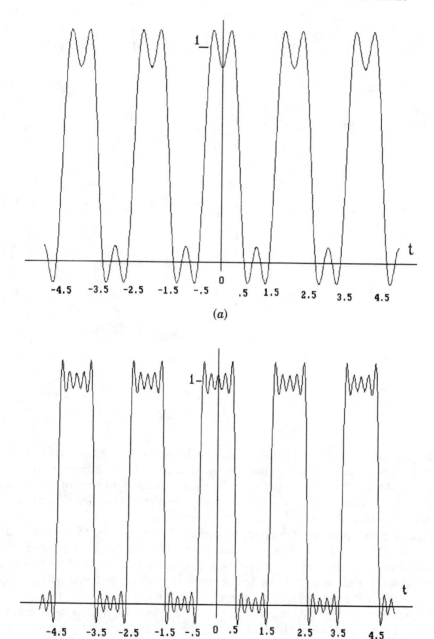

(a)

(b)

Figure 1.4 Fourier series approximation to the boxcar function. (a) Three terms, (b) nine terms, (c) 30 terms.

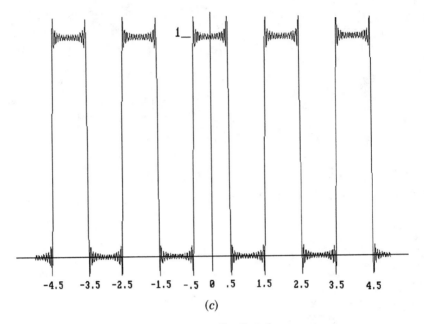

-4.5   -3.5   -2.5   -1.5   -.5   0   .5   1.5   2.5   3.5   4.5

(*c*)

Figure 1.4   *(Continued)*

boxcar function. We note that to accurately determine if the Fourier series converges to the generating function we must consider an infinite number of terms. As it turns out in this case, even with an infinite number of terms (not feasible to show) the series does converge to the boxcar function at every point except at the edges.

In the two previous examples, we were able to perform the integral calculations required to obtain analytical expressions for the Fourier series coefficients. Anyone who can recall basic calculus will appreciate the fact that integration is often a challenging operation to perform and that many seemingly innocent integral expressions are difficult, if not impossible, to evaluate. Thus, when dealing with the Fourier series we have the following concerns. First, what conditions must be placed on the function that will guarantee that the integral expressions for the Fourier series coefficients exist? Second, if it can be determined that they do indeed exist, is it possible for us to analytically evaluate them using known integration techniques? Third, if they exist and we can evaluate them, are we guaranteed that when we add up the terms of the series using these coefficients, they will converge to the generating function? These are the topics that are discussed in Chapter 5.

THE FOURIER TRANSFORM

Given a function $f(t)$, we define the Fourier transform pair as

$$F(w) = \int_{-\infty}^{\infty} f(t)e^{-2\pi iwt}\,dt, \tag{1.5}$$

$$f(t) = \int_{-\infty}^{\infty} F(w)e^{2\pi iwt}\,dw. \tag{1.6}$$

Equation (1.5) is known as the *(direct) Fourier transform*, and equation (1.6) is known as the *inverse Fourier transform*. We say that $F(w)$ is the Fourier transform of $f(t)$ and that $f(t)$ is the inverse Fourier transform of $F(w)$. Notationally, we write

$$F(w) = \mathcal{F}[f(t)] \quad \text{and} \quad f(t) = \mathcal{F}^{-1}[F(w)]$$

to denote the fact that $f(t)$ and $F(w)$ are Fourier transform pairs. As can be seen from equations (1.5) and (1.6), the Fourier transform may be considered a transformation that maps the function $f(t)$ to another function $F(w)$. Similarly, the inverse Fourier transform maps $F(w)$ back to $f(t)$.

As our first example, let us calculate the Fourier transform of the one-sided exponential function defined as

$$f(t) = \begin{cases} e^{-at}, & t \geq 0, \\ 0, & \text{otherwise.} \end{cases}$$

This function is illustrated graphically in Figure 1.5. Using equation (1.5) we calculate its Fourier transform as follows:

$$F(w) = \int_{-\infty}^{\infty} f(t)e^{-2\pi iwt}\,dt = \int_{0}^{\infty} e^{-at}e^{-2\pi iwt}\,dt,$$

$$F(w) = \int_{0}^{\infty} e^{-(a+2\pi iw)t}\,dt = \left. \frac{-1}{a+2\pi iw}e^{-(a+2\pi iwt)} \right|_{0}^{\infty},$$

$$F(w) = \frac{1}{a+2\pi iw} = \frac{a-2\pi iw}{a^2+4\pi^2 w^2}.$$

This function is obviously complex valued with a real and imaginary portion. The real portion is plotted in Figure 1.6a, and the imaginary portion is plotted in Figure 1.6b.

Complex functions can also be represented in the complex exponential form, that is,

$$F(w) = \|F(w)\| e^{i\varphi(w)},$$

where $\|F(w)\|$ represents the amplitude of the function, and $\varphi(w)$ represents the phase angle. For the transform of the exponential function that we are currently dealing with we have

$$\|F(w)\| = \frac{1}{(a^2+4\pi^2 w^2)^{1/2}} \quad \text{and} \quad \varphi(w) = \tan^{-1}\left[\frac{-2\pi w}{a}\right].$$

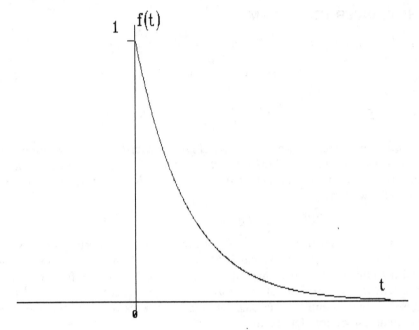

Figure 1.5 The exponential function.

This complex exponential form is graphically illustrated in Figure 1.7.

Under the most general circumstances, the Fourier transform of a real function (or signal) will be a complex function (such as in the case of the previous example). However, now let us consider a special case as our next example, in which the Fourier transform also turns out to be a real function. The function that we transform is the *pulse* or *tophat function*, which is mathematically defined as:

$$p_a(t) = \begin{cases} 1, & -a \leq t \leq a, \\ 0, & \text{otherwise.} \end{cases}$$

This function is shown in Figure 1.8. Again, the Fourier transform of this function is obtained by direct application of equation (1.5), that is,

$$F(w) = \int_{-\infty}^{\infty} p_a(t)e^{-2\pi iwt}\, dt = \int_{-a}^{a} e^{-2\pi iwt}\, dt = \left.\frac{e^{-2\pi iwt}}{-2\pi iw}\right|_{-a}^{a},$$

$$F(w) = \frac{e^{+2\pi iwa} - e^{-2\pi iwa}}{2\pi iw} = \frac{\sin(2\pi wa)}{\pi w},$$

$$F(w) = \frac{2a\sin(2\pi wa)}{2\pi wa} = 2a\,\text{sinc}(2\pi wa).$$

This transform is shown in Figure 1.9. As can be seen, the Fourier transform of the pulse function is indeed a real-valued function. In other words, it has

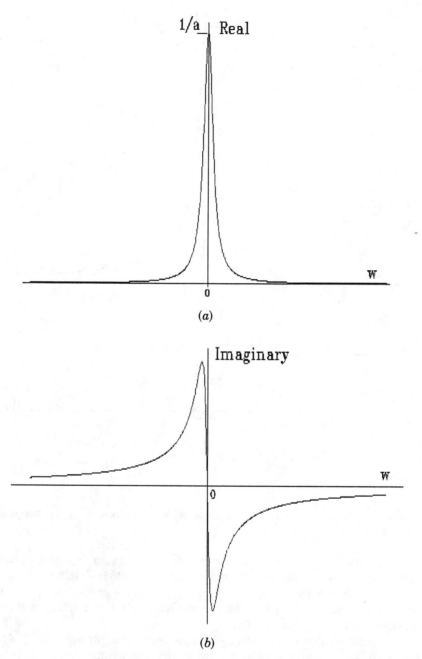

Figure 1.6 Fourier transform of exponential function. (a) Real, (b) imaginary.

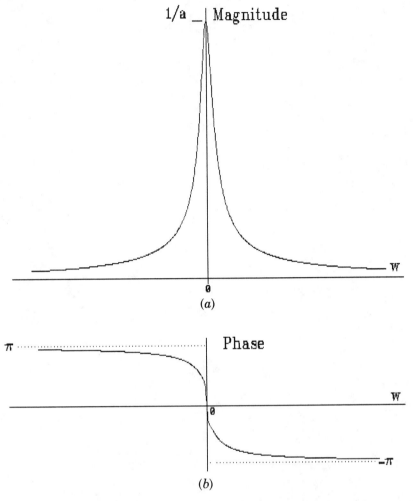

Figure 1.7 Fourier transform of exponential function—complex form. (a) Magnitude, (b) phase.

no imaginary portion. Real functions can also be placed in the complex exponential, or amplitude and phase, format. The amplitude is simply the absolute value of the real portion, and the phase is either 0 or π depending upon if the function is positive or negative. Shown in Figure 1.10 is the amplitude–phase representation of the sinc function.

The question that now emerges is, if we were to apply equation (1.6) to the above two example results, will we obtain the original function? This is an extremely important concept and one that everyone assumes to be correct when manipulating the Fourier transform. Texts are filled with the calculation of the Fourier transform of various functions, but very seldom does one see equation (1.6) applied to determine the inverse Fourier transform.

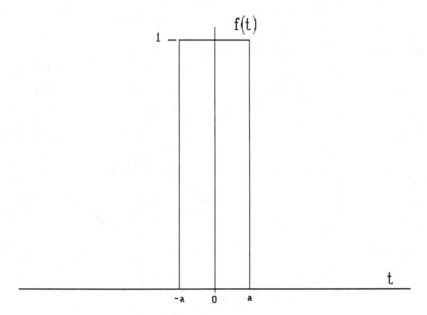

Figure 1.8   The pulse function.

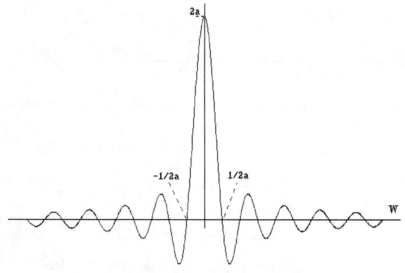

Figure 1.9   Fourier transform of pulse function.

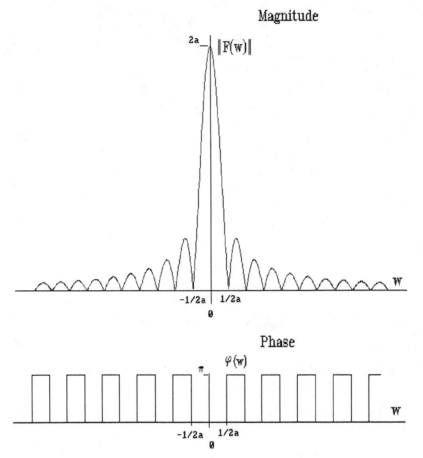

Figure 1.10 Fourier transform of pulse function—amplitude and phase (polar) representation.

As an example, let us use the inverse Fourier transform to show the following result:

$$\int_{-\infty}^{\infty} \text{sinc}(2\pi ax)\,dx = \frac{1}{2a}.$$

This is accomplished by making the assumption that equation (1.6) can indeed be used to recover the pulse function from the sinc function, that is,

$$p_a(t) = \mathcal{F}^{-1}[2a\,\text{sinc}(2\pi wa)] = 2a\int_{-\infty}^{\infty}\text{sinc}(2\pi wa)e^{2\pi iwt}\,dw.$$

Now, in the above equation, we set $t = 0$ and use the fact that $p_a(0) = 1$ to obtain our desired result, that is,

$$1 = 2a\int_{-\infty}^{\infty}\text{sinc}(2\pi ax)\,dx.$$

In the above derivation we made the assumption that the Fourier transform of a function was unique. In other words, the only possible inverse transform of $F(w) = 2a\,\text{sinc}(2\pi wa)$ is the pulse function $p_a(t)$.

Another issue that we should be concerned with is the existence of the integrals involved with the calculation of the Fourier transform. As it turns out, when we examine the question of existence of the Fourier integrals, we will also be exploring the uniqueness of the Fourier transform. These topics are discussed in Chapter 6.

THE IMPULSE FUNCTION

Physically, an *impulse* is considered to be an extremely brief, very intense, unit area pulse. Mathematically, the impulse has zero width and infinite height. While such a true impulse function does not exist in the real world, it is an extremely powerful mathematical, or theoretical, tool used to study the behavior of many phenomena. For example, in physics an impulse function is used to discuss such quantities as point charges, point masses, point sources, and concentrated forces. This impulse function is the principal tool used to examine the behavior of dynamic, as well as optical, systems. An impulse function is also known as a *delta* or *Dirac delta* function. The impulse function can also be manipulated to obtain other interesting mathematical Fourier results.

There are several ways in which an impulse function can be defined. Without the use of distribution theory (which is presented in Chapter 4) we must be satisfied with describing them as the limiting case of a sequence of other functions. There are several functions that can be used as the generating sequence, such as a sinc function, a pulse function, or a Gaussian function. In the derivation that follows we use a sequence of pulse functions. Let us consider the sequence of pulse functions $\{p_{1/n}(t)\}$ defined as

$$p_{1/n}(t) = \begin{cases} n/2, & |t| \le 1/n, \\ 0, & \text{otherwise.} \end{cases}$$

A term of this sequence is shown in Figure 1.11. We note that the area of this pulse is $(n/2)(2/n) = 1$. Mathematically, we write

$$\int_{-\infty}^{\infty} p_{1/n}(t)\,dt = \int_{-1/n}^{1/n} \frac{n}{2}\,dt = 1 \qquad \text{for all values of } n.$$

We now define the impulse, or delta function, as the limit (as n approaches ∞) of this sequence of unit area impulse functions, that is,

$$\delta(t) = \lim_{n \to \infty} p_{1/n}(t).$$

This limiting process is illustrated in Figure 1.12 for $n = 1, 2, 3, 4$. From the figure we see that as n increases, the width of each pulse decreases toward zero, whereas the height, or intensity, increases without bound. However, the

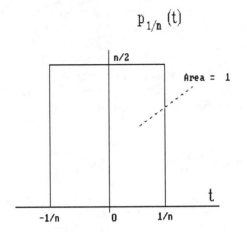

Figure 1.11 Generating pulse function.

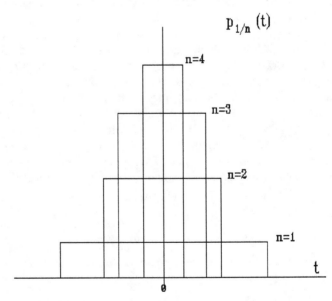

Figure 1.12 Definition of impulse function as limit of a sequence of pulse functions of constant area.

area (or strength) of each pulse remains equal to unity. Thus, it seems reasonable to write

$$\int_{-\infty}^{\infty} \delta(t)\,dt = \int_{-\infty}^{\infty} \lim_{n\to\infty} p_{1/n}(t)\,dt = \lim_{n\to\infty} \int_{-\infty}^{\infty} p_{1/n}(t)\,dt = 1.$$

In the previous section we showed that the Fourier transform of a unit pulse function of half-width a was given as $2a\,\text{sinc}(2\pi wa)$. Thus, for each term of the sequence of pulse functions of half-width $1/n$ (and magnitude $n/2$) we know that its Fourier transform is given as $\text{sinc}(2\pi w/n)$. From this we deduce

$$\mathcal{F}[\delta(t)] = \mathcal{F}\left[\lim_{n\to\infty} p_{1/n}(t)\right] = \lim_{n\to\infty} \mathcal{F}[p_{1/n}(t)]$$

$$= \lim_{n\to\infty} \frac{\sin(2\pi w/n)}{2\pi w/n} = 1.$$

Thus, in a very crude mathematical sense we have "demonstrated" that the Fourier transform of the impulse function is the constant function $F(w) = 1$. While this approach may be somewhat intuitively gratifying, it is by no means a justifiable mathematical procedure. The biggest objection to our previous line of reasoning was the interchange of the infinite limit and integration process (required by the Fourier transform operation). To properly appreciate the nature and validity of the delta function it is necessary to approach it from a distribution theory point of view. This is discussed in Chapters 4 and 7.

SUMMARY

The purpose of this chapter was to provide the justification, or motivation, for looking at the analytical behavior of the Fourier series and Fourier transform. For the Fourier series we must consider what conditions should be placed on the function $f(t)$ such that the integral expressions of equations (1.2) and (1.3) exist and can be integrated to obtain the Fourier series coefficients A_k and B_k. We must also study the circumstances under which the infinite summation of equation (1.1) will converge to the generating function $f(t)$.

For the Fourier transform we are also interested in discovering what conditions the function $f(t)$ must satisfy in order to guarantee that its Fourier transform, as given by equation (1.5), will exist. In addition to this, we are also interested in examining the uniqueness of the Fourier transform. In other words, when will equation (1.6) (the inverse Fourier transform) return $f(t)$ from $F(w)$ uniquely? Finally, in this chapter we saw that more study is required to properly define and characterize the delta functon and its Fourier transform.

PROBLEMS

1 For the triangle sawtooth function shown in Figure 1.1, apply equation (1.3) to show that $B_k = 0$ for all k.

2 For the boxcar function shown in Figure 1.3, apply equation (1.3) to show that $B_k = 0$ for all k.

3 Consider the triangle sawtooth function defined over the interval $[-2,2]$ as

$$f(t) = \begin{cases} 1+t, & t \in [-1,0], \\ 1-t, & t \in (0,1], \\ 0, & \text{otherwise.} \end{cases}$$

In this case consider the function to be periodic over the interval $[-2,2]$ and calculate its Fourier series coefficients.

4 Consider the boxcar function defined over the interval $[-2,2]$ as

$$f(t) = \begin{cases} 1, & t \in [-\frac{1}{2},\frac{1}{2}], \\ 0, & \text{otherwise.} \end{cases}$$

In this case consider the function to be periodic over the interval $[-2,2]$ and calculate its Fourier series coefficients.

5 Consider the ramp sawtooth function defined as $f(t) = t$ over the interval $[-1,1]$. Consider this function to be periodic over the same interval $[-1,1]$ and calculate its Fourier series coefficients.

6 Show that

$$\frac{1}{a+2\pi iw} = \frac{a-2\pi iw}{a^2+4\pi^2w^2}.$$

(*Hint:* Multiply both numerator and denominator by the complex conjugate of the denominator.)

7 Using equation (1.6) for the inverse Fourier transform, show

$$\int_{-\infty}^{\infty} \text{sinc}(2\pi ax)e^{i\pi x/4}\,dx = \frac{1}{2a}.$$

8 Using equation (1.6) for the inverse Fourier transform, show

$$\int_{-\infty}^{\infty} \text{sinc}(2\pi ax)e^{i2\pi x}\,dx = 0.$$

BIBLIOGRAPHY

Weaver, H. J., *Applications of Discrete and Continuous Fourier Analysis*, Jown Wiley & Sons, New York, 1983.

Cooley, J. W. and J. W. Tukey, "An Algorithm for the Machine Calculation of Complex Fourier Series," *Mathematical Computations*, **19**, April 1965.

2

BASIC MATHEMATICAL BACKGROUND

In the previous chapter we furnished several example calculations of both the Fourier transform and the Fourier series representation of a function. The purpose of that presentation was to provide the motivation for studying the more theoretical aspects of Fourier analysis. To accomplish such a theoretical study we require a reasonable understanding of integration theory as well as the concept of convergence of both functions and series. Obviously, the study of convergence is germane to the understanding of the Fourier series representation of a function. However, it is also required to understand the development of integration theory presented in Chapter 3.

In this chapter we present the basic mathematical terminology and concepts that are necessary to study convergence of functions, sequences, and series. The material covered in this chapter also serves as a foundation for the remaining chapters in this text.

FUNDAMENTAL SET THEORY

We define a *set* as any collection of objects. The objects, or elements, of the set may be anything from apples to zebras. There is nothing abstract or mysterious implied by our definition of a set. It means exactly what it says and conforms to our intuitive concept of a set, such as a set of books or a set of tools. We usually denote sets by capital letters and denote the elements of the set by lowercase letters. Notationally, we write $x \in A$ to mean x is an element of the set A and write $x \notin A$ to mean x is not an element of the set A.

In general, there need not be any specific relationship or common property among the elements of a set. For example, we could define a set A as the collection of all automobiles and dogs in Chicago. However, when there is the possibility of placing some arrangement or relationship among the elements of a set we usually refer to the set as a *space*. In reality, however, a space is just another name for a set.

A set is called a *finite set* if it contains a finite number of elements and is called an *infinite set* if it contains an infinite number of terms. For example, the set A, consisting of all integers from 1 to 5, notationally written as $\{1,2,3,4,5\}$, is a finite set. The set B, consisting of all integers greater than 0, is an example of an infinite set.

If A is a set, then a set B is called a *subset* of A if every element of B is also an element of A. Notationally, we write $B \subset A$ or $A \supset B$ to indicate that B is a subset of A. We sometimes say that A contains B or that A covers B. For example, if A is the set given by $\{1,2,3,4,5\}$ and B is the set given by $\{1,4\}$, then clearly $B \subset A$. Note, by this definition, that $A \subset A$ because every element of A is obviously an element of A. Two sets, A and B, are called *equal* if $A \subset B$ and $B \subset A$. We can easily see, by applying our definition of a subset, that equality of A and B implies that both sets contain the exact same elements. B is called a *proper subset* of A if $B \subset A$ but B is not equal to A, or, in other words, A contains B but B does not contain A.

A set that has no elements is called the *empty set* and is denoted as $\{ \ \}$.

SET ALGEBRA

We now consider some *set algebra*, which are basically rules of arithmetic defined for, and applied to, sets. These rules, known as the union and intersection, are quite similar to those of addition and multiplication defined for the real numbers.

The *union* (sometimes called the *sum*) of two sets A and B, written as $A \cup B$, is defined as the set consisting of the objects that are elements of *either* A or B.

The *intersection* (sometimes called the *product*) of two sets A and B, written as $A \cap B$, is defined as the set consisting of the objects that are elements of *both* A and B. To illustrate the union and intersection of sets, consider the following four sets of integers:

$$A = \{1,2,3,4,5,6,7\},$$
$$B = \{3,4,5,6,7,8\},$$
$$C = \{4,5\},$$
$$D = \{10,11,12,13,14\}.$$

Now by simply applying the above two definitions we obtain the following:

$$A \cup B = \{1,2,3,4,5,6,7,8\},$$

$$A \cup C = \{1,2,3,4,5,6,7\} = A,$$

$$A \cup D = \{1,2,3,4,5,6,7,10,11,12,13,14\},$$

$$A \cap B = \{3,4,5,6,7\},$$

$$A \cap C = \{4,5\},$$

$$C \cap D = \{\ \},$$

$$C \cap C = \{4,5\}.$$

The union and intersection can be applied to any number of sets. That is, the union of the sets $A_1, A_2, \ldots, A_n$, written as

$$\bigcup_{i=1}^{n} A_i,$$

is defined as the set consisting of elements which are elements of A_1 or A_2 or ...or A_n. Thus, if x is an element of $\cup A_i$, then x must be an element of *at least one* of the sets $A_i (i = 1, \ldots, n)$.

Similarly, the intersection of the sets $A_1, A_2, \ldots, A_n$, written as

$$\bigcap_{i=1}^{n} A_i,$$

is defined as the set consisting of elements which are elements of A_1 and A_2 and...and A_n. Thus, if x is an element of $\cap A_i$, then x must be an element of *every* set $A_i (i = 1, \ldots, n)$. It is important to note that n may be allowed to go to infinity. Using the previously defined four sets of integers, we have

$$A \cup B \cup D = \{1,2,3,4,5,6,7,8,10,11,12,13,14\},$$

$$A \cap B \cap C = \{4,5\}.$$

By a careful choice of order (use of parenthesis) we can combine both the union and intersection of sets. For example,

$$(A \cup B) \cap C = \{1,2,3,4,5,6,7,8\} \cap \{4,5\} = \{4,5\},$$

$$A \cup (B \cap C) = \{1,2,3,4,5,6,7\} \cup \{4,5\} = \{1,2,3,4,5,6,7\},$$

$$(A \cap B) \cup C = \{3,4,5,6,7\} \cup \{4,5\} = \{3,4,5,6,7\},$$

$$A \cap (B \cup C) = \{1,2,3,4,5,6,7\} \cap \{3,4,5,6,7,8\} = \{3,4,5,6,7\}.$$

The *complement of A relative to B*, written as $B - A$, is defined as the set consisting of those elements of B which are not elements of A (i.e., $x \in B$,

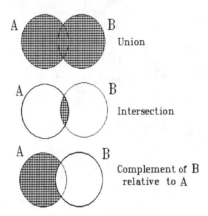

A   B

Union

A   B

Intersection

A   B

Complement of B
relative to A

Figure 2.1 Set algebra.

$x \notin A$). When applied to our previous examples, we find

$$B - A = \{8\},$$
$$A - B = \{1,2\},$$
$$D - C = \{10,11,12,13,14\},$$
$$A - A = \{\ \},$$
$$C - D = \{4,5\}.$$

It is sometimes helpful to visualize these new sets (union, intersection, and complement) by drawing the sets A and B as circles in a two-dimensional space as shown in Figure 2.1. These figures are also referred to as *Venn diagrams*.

Two sets A and B are called *disjoint* if their intersection is the empty set. In our previous examples we illustrated that the sets C and D are disjoint. Using Venn diagrams, disjoint sets would appear as shown in Figure 2.2.

Thus far, we have said nothing about any order or special arrangement of the elements in a set. We now consider a very special set, of only two elements, in which order is important. An *ordered pair* is a set, consisting of two elements a and b, such that (a,b) is not necessarily equal to (b,a). A familiar example of an ordered pair is a complex number z written as the ordered pair (x,y), in which x is the real part and y is the imaginary part. For example, the complex number with real component 2.3 and imaginary component 4.01 is given by

$$z = \{2.3, 4.01\}.$$

Having defined what we mean by an ordered pair, we are now able to define the Cartesian product of two sets A and B. Given two sets, A and B, the set of all ordered pairs (a,b) such that $a \in A$ and $b \in B$ is called the *Cartesian product* of A and B and is denoted by $A \times B$. Here we see that each

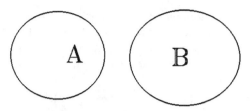

Figure 2.2 Two disjoint sets.

element of $A \times B$ is a two-dimensional ordered element or, in other words, each element of $A \times B$ is itself a set. Thus, $A \times B$ is a collection of sets or a set of sets. This Cartesian product set (or simply product set) will be useful when discussing the concept of a function.

LAWS OF INTERNAL COMPOSITION

The union and intersection of sets are examples of set algebra, which are rules of arithmetic applied to two or more sets. We now discuss the algebra of the elements of a set, or the rules of arithmetic applied to the individual elements of a set. These rules are called *laws of internal composition*. When we talk about these laws we necessarily limit the types of elements that a set can contain. For example, it doesn't make much sense to consider the product, or sum, of the elements of the set that is a collection of alphabetical characters. Therefore, from here on we limit our attention to sets consisting of elements that are mathematical in nature, such as integers, real numbers, matrices, continuous functions, and so on.

Let us first assume we are given a set of elements E. A *law of internal composition* (such as the sum and product) is a rule that combines any two elements, of the set E, to yield a third element *that is also an element of the set E*. For example, if we consider E to be the set of all integers, then addition of integers is a law of internal composition for this set, since the sum of any two integers is also an integer and, consequently, an element of E. As a counterexample, let us consider the finite set of integers K given by $\{1,2,3,4\}$. In this case, addition of integers does not form a law of internal composition, because, in general, the sum of any two integers in this set will not be an element of the set. For example, the sum of 2 and 3 is equal to 5, which is not an element of K.

Just as we had the sum (union) and product (intersection) as our arithmetic rules when we dealt with set algebra, we will deal with the sum (addition) and product (multiplication) when considering laws of internal composition. These rules, or operations, must be carefully defined for each set to which they are applied. When the elements of the set are integers or real numbers, the laws are straightforward and quite familiar to us. We use the symbol "+" for addition and "·" for multiplication, with the understanding that their definitions depend upon the type of elements in the set to which they are being applied. Also dependent upon the set for which these laws are defined are the properties of associativity, commutativity, and distributivity.

We say the addition law $(+)$ is *associative* if, for every x, y, z in the set A, we have

$$(x + y) + z = x + (y + z).$$

Similarly, the product law $(\cdot)$ is *associative* if, for every x, y, z in A, we have

$$(x \cdot y) \cdot z = x \cdot (y \cdot z).$$

The standard addition and product laws defined for integers are associative. For example,

$$(5 + 3) + 7 = 8 + 7 = 5 + (3 + 7) = 5 + 10 = 15$$

and

$$(5 \cdot 3) \cdot 7 = (15) \cdot 7 = 5 \cdot (3 \cdot 7) = 5 \cdot 21 = 105.$$

We say the addition law $(+)$ is *commutative* if, for every $x, y \in A$, we have $x + y = y + x$. Similarly, for the product law $(\cdot)$, we have commutativity if $x \cdot y = y \cdot x$. When dealing with integers and/or real numbers the commutative property seems almost trivial. For example,

$$5 + 2 = 2 + 5 = 7 \quad \text{and} \quad 5 \cdot 2 = 2 \cdot 5 = 10.$$

However, we point out that these properties do not always exist. For example, if the elements of the set A are matrices and the product law, for this set, is the usual definition of matrix multiplication, then $X \cdot Y$ is not necessarily equal to $Y \cdot X$. For example,

$$\begin{bmatrix} 3 & 4 \\ 1 & 8 \end{bmatrix} \begin{bmatrix} 5 & 2 \\ 6 & 7 \end{bmatrix} = \begin{bmatrix} 39 & 34 \\ 53 & 58 \end{bmatrix}$$

but

$$\begin{bmatrix} 5 & 2 \\ 6 & 7 \end{bmatrix} \begin{bmatrix} 3 & 4 \\ 1 & 8 \end{bmatrix} = \begin{bmatrix} 17 & 36 \\ 25 & 80 \end{bmatrix}.$$

Thus we see that the existence of these properties depends upon the set to which the laws are applied and also upon how they are defined.

Thus far we have considered properties that applied to the sum and product laws individually. We now consider a property, known as *distributivity*, that is concerned with a combination of the two laws. The product law $(\cdot)$ is said to be *distributive* with respect to the addition law $(+)$ if, for any $x, y, z \in A$, we have

$$(x + y) \cdot z = (x \cdot z) + (y \cdot z) \quad \text{and} \quad z \cdot (x + y) = (z \cdot x) + (z \cdot y).$$

Again, for the set of integers we offer the following example:

$$(3 + 5) \cdot 4 = 8 \cdot 4 = 3 \cdot 4 + 5 \cdot 4 = 32$$

and

$$3 \cdot (5 + 4) = 3 \cdot 9 = 3 \cdot 5 + 3 \cdot 4 = 27.$$

SPECIAL ELEMENTS OF A SET

We now consider two special elements of a set known as the *unit element* and the *inverse element*. We note here that it only makes sense to talk about these elements when a law of internal composition is defined on the set. The *unit element under addition* is an element of a set A, usually denoted as 0, such that for every $x \in A$ we have

$$x + 0 = 0 + x = x.$$

Similarly, the *unit element under multiplication* is an element of a set A, usually denoted as 1, such that

$$x \cdot 1 = 1 \cdot x = x.$$

For the real number system we obviously have 0 as the unit element under addition and have 1 as the unit element under multiplication. For the complex number system the unit element under addition is $(0,0)$ and the unit element under multiplication is $(1,0)$. For the set of all 3×3 matrices the unit element under addition is

$$[0] = \begin{bmatrix} 0 & 0 & 0 \\ 0 & 0 & 0 \\ 0 & 0 & 0 \end{bmatrix},$$

while the unit element under multiplication is

$$[1] = \begin{bmatrix} 1 & 0 & 0 \\ 0 & 1 & 0 \\ 0 & 0 & 1 \end{bmatrix}.$$

An element x, in the set A, has an *inverse under addition* if there exists another element $\underline{x}'$, also in A, such that

$$x + \underline{x}' = \underline{x}' + x = 0.$$

Similarly, an element x in the set A has an *inverse under multiplication* if there exists another element x', also in A, such that

$$x \cdot x' = x' \cdot x = 1.$$

For the set of real numbers, given any element a, the inverse under addition is simply the negative of the element $-a$ and the inverse under multiplication is the reciprocal of a, namely, $1/a$. Given any complex number (x,y), the inverse under addition is given as the element $(-x,-y)$ since

$$(x,y) + (-x,-y) = (x-x, y-y) = (0,0) = \mathbf{0}.$$

The inverse under multiplication is given as the element

$$z^{-1} = \left(\frac{x}{x^2 + y^2}, \frac{-y}{x^2 + y^2} \right)$$

because

$$zz^{-1} = (x,y)\left(\frac{x}{x^2 + y^2}, \frac{-y}{x^2 + y^2} \right),$$

$$zz^{-1} = \left(\frac{x^2 + y^2}{x^2 + y^2}, \frac{x(-y) + yx}{x^2 + y^2} \right) = (1,0).$$

A set may possess an inverse under one law but not under the other. For example, the set of all integers $\mathbf{I}$ has an inverse under addition, since given any integer $n \in I$, the inverse is simply given by $-n \in I$, where $n + (-n) = 0$. On the other hand, the set of all integers has no inverse under multiplication, since given any $n \in I$, the inverse element would have to be $1/n$ because $n(1/n) = 1$. However, in general, $1/n$ is not an element of the set of integers.

GROUPS, FIELDS, AND VECTOR SPACES

Having defined what we mean by a set, and also having defined an algebra between the elements of the set, we can now consider a few special sets that have one or more of the algebraic laws defined on them. The first special case we consider is a group which is a set with one internal law defined on it and satisfying a few simple conditions. More specifically, a set A for which one internal law of composition is defined is called a *group* if that law satisfies the following properties:

(1) The law is associative.
(2) The set A possesses a unit element under that law.
(3) For every element x of A, there exists an inverse element x' (also in A) under that law.

If, in addition to the above conditions, the law is also commutative, then the set is called an *Abelian group*.

The set that consists of all integers is an Abelian group under the law of addition since it satisfies all the conditions as stated above. We note, however, that the set of all integers is not a group under the law of multiplication because, as we have already seen, there is no inverse element under multiplication. The set of all complex numbers forms an Abelian group under addition since:

(1) Addition is associative.
(2) The unit element exists and is equal to $(0,0)$.
(3) The inverse element exists and is equal to $(-x,-y)$.
(4) Addition is commutative.

We have just seen that an Abelian group is simply a set that has a law of internal composition that satisfies certain properties. If we now add a second law to an Abelian group, we obtain what is known as a *field*.

A set A is called a *field* if it forms an Abelian group under the addition law and, in addition, has a product law defined on it which satisfies the following conditions:

(a) Multiplication is associative.
(b) The set A possesses a unit element under the product law.
(c) Multiplication is commutative.
(d) For every element x of A, except 0 (the unit element under addition), there exists an inverse element under the product law.
(e) The product law is distributive with respect to the addition law.

We note that conditions (a)–(d) are almost the requirements that the set A form an Abelian group under multiplication. The only exception is condition (d), which excludes 0, the unit element under addition. This is obviously necessary for such fields as the real number system since 0 has no inverse (or reciprocal) under multiplication. Condition (e) relates the two laws.

In summary, a set A is called a field if there are two laws [addition (+) and multiplication ($\cdot$)] defined on A such that for every $x, y, z \in A$ we have:

(1) $x + (y + z) = (x + y) + z$.
(2) $x + 0 = 0 + x = x$ for all $x \in A$.
(3) $x + (-x) = (-x) + x = 0$.
(4) $x + y = y + x$.

(a) $x \cdot (y \cdot z) = (x \cdot y) \cdot z$.

(b) $x \cdot 1 = 1 \cdot x = x$.

(c) $x \cdot y = y \cdot x$.

(d) $x \cdot x' = x' \cdot x = 1$ for all $x \in A$, $x \neq 0$.

(e) $(x + y) \cdot z = (x \cdot z) + (y \cdot z)$ and $z \cdot (x + y) = (z \cdot x) + (z \cdot y)$.

The set of complex numbers forms a field because for this set we are able to define a sum and a product law such that (see problem 2):

(1) Addition is associative.

(2) $(0,0)$ is the unit element under addition.

(3) $(-x,-y)$ serves as the inverse element under addition.

(4) Addition of complex numbers is commutative.

(a) Multiplication of complex numbers is associative.

(b) $(1,0)$ is the unit element under multiplication.

(c) Multiplication of complex numbers is commutative.

(d) For any complex number, except $(0,0)$, we can define an inverse under multiplication as

$$z^{-1} = \left(\frac{x}{x^2 + y^2}, \frac{-y}{x^2 + y^2} \right).$$

(e) For complex numbers, multiplication is distributive with respect to addition.

There is another special set, known as a *vector space*, that is deserving of our attention. A vector space is a rather interesting combination of two sets V and F. The set V is called the *vector space*, and the set F is called a *field*. The elements of V are known as *vectors*, and the elements of F are called *scalars*. In a vector space, we first define an internal law of addition among the elements of V and require that V be an Abelian group under this law. Whereas for a field we added a second *internal* law of multiplication, for a vector space we add a second *external* law of multiplication which forms the product of an element of V with an element of F and yields a third element, which belongs to the set V. More formally, we have the following definition. Given a field F and a set V, V is called a *vector space* if it satisfies the following conditions:

(A) V forms an Abelian group under an internal law of addition. That is, for any $x, y, z \in V$:

(1) Addition is associative, that is,

$$(x + y) + z = x + (y + z).$$

(2) There exists a unit element (0) under addition such that

$$x + 0 = 0 + x = x \qquad \text{for all} \quad x \in V.$$

(3) There exists an inverse element for all $x \in V$, such that

$$x + (-x) = (-x) + x = 0.$$

(4) Addition is commutative, that is,

$$x + y = y + x.$$

(B) To every pair $a \in F$ and $x \in V$, there corresponds an element $y = ax \in V$, called the product of y and x, such that:
(1) Multiplication by scalars is associative, that is,

$$a(bx) = (ab)x \qquad \text{for all} \quad a, b \in F, \quad x \in V.$$

(2) V has a unit element under multiplication (denoted as 1) which is also the unit element under multiplication in F, that is,

$$1x = x \qquad \text{for all} \quad x \in V.$$

(C) Multiplication by scalars is distributive with respect to vector addition, that is, for $a \in F$ and $x, y \in V$, we have

$$a(x + y) = (ax) + (ay),$$
$$(x + y)a = (xa) + (ya).$$

(D) Multiplication by vectors is distributive with respect to scalar addition, that is, for $a, b \in F$ and $x \in V$, we have

$$(a + b)x = (ax) + (bx),$$
$$x(a + b) = (xa) + (xb).$$

In general terminology we say **V** is a *vector space over the field F*.

Probably the most familiar example of a vector space is the set of complex numbers over the real field. In this situation we define multiplication of any scalar $a \in R^1$ and any vector, or complex number, $z \in C$ in the following manner:

$$az = a(x, y) = (ax, ay).$$

We now proceed to show that the set of complex numbers C does indeed form a vector space over field of real numbers R^1.

(A) In the previous example we established that the set of complex numbers formed an Abelian group under addition.
(B) For any scalars $a, b \in R^1$ and complex number $z \in C$ we have

(1) $a(bz) = a(b(x,y)) = a(bx,by) = (abx,aby) = ab(x,y) = (ab)z$.

(2) The unit element under multipliction in C is $(1,0)$, which is simply the real number 1.

(C) For any $a,b \in R^1$ and $z_1, z_2 \in C$ we have

$$a(z_1 + z_2) = a(x_1 + x_2, y_1 + y_2) = (ax_1 + ax_2, ay_1 + ay_2)$$
$$= (ax_1, ay_1) + (ax_2, ay_2) = az_1 + az_2$$
$$(a + b)z_1 = (a + b)(x_1, y_1) = ((a + b)x_1, (a + b)y_1)$$
$$= (ax_1 + bx_1, ay_1 + by_1)$$
$$= (ax_1, ay_1) + (bx_1, by_1)$$
$$= a(x_1, y_1) + b(x_1, y_1) = az_1 + bz_1.$$

The commutative properties of addition for both R^1 and C establish the other condition required by (D).

A somewhat more abstract example of a vector space is the set **V** of polynomials, with real coefficients, in the variable x. In this case, the field is the set of real numbers. Addition is defined by the ordinary rules for adding two polynomials together, and the unit element under addition is the polynomial equal to 0 for all x.

FUNCTIONS, RELATIONS, AND SEQUENCES

The concept of a function is crucial to the study of Fourier analysis. When describing the Fourier series representation of a function, we place certain restrictions on the function that will guarantee the existence and convergence of its series representation. When dealing with the Fourier transforms, we also place conditions on the function that will ensure that it possesses a Fourier transform and we define the circumstances under which the transform is unique. In both situations we are dealing with the concept of a function. Therefore, in this section we present a proper and general definition of a function. We begin our presentation with the defintion of a relation, of which a function is a special case. Given two sets A and B, we define a *relation* S each time we set up a correspondence, or mapping, between every $x \in A$ and some $y \in B$. A *function* is a special relation in which there can only be one $y \in B$ associated with each $x \in A$. We illustrate these concepts with the following examples: Let A be the set of integers $\{1,2,3,4,5\}$, and let B be the set of letters $\{a,b,c,d,e,f,g\}$. In our first example we define a relation S between A and B as: $1 \to b$, $2 \to c$, $3 \to a$, $4 \to e$, $5 \to e$. This relation is illustrated graphically in Figure 2.3, where the dashed lines denote the correspondence between the various elements from the two sets. A relation can be

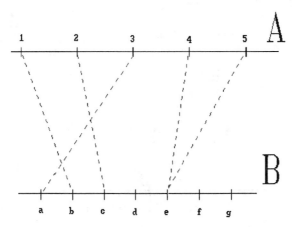

Figure 2.3 Graphical representation of a relation that is also a function.

written more compactly as ordered pairs (x,y), where $x \in A$ and $y \in B$. Thus, we write the above relation as the set

$$S = \{(1,b),(2,c),(3,a),(4,e),(5,e)\}.$$

Therefore, we see that a relation (or a function) from A to B can be defined as the set of ordered pairs (x,y) such that $x \in A$ and $y \in B$. Notationally, we sometimes write $y = S(x)$ to mean $(x,y) \in S$. In the above example the relation S is also a function because for each $x \in A$ there is only one $y \in B$. Note, however, that the same y may be mapped from different values of x (e.g., $e = S(4)$ and $e = S(5)$). As an example of a relation that is not a function, we have

$$S = \{(1,a),(2,b),(3,c),(4,d),(5,e),(1,b)\}.$$

In this case we see that the element $x = 1$ is mapped to both elements a and b of the set B. Therefore, S is not a function. This relation is graphically illustrated in Figure 2.4.

We now limit our attention exclusively to relations that are functions. When dealing with functions we define two special sets known as the *domain* and the *range* of the function. Formally, we have: Let f be a function from A to B, that is, let f be the set of ordered pairs (x,y), where $x \in A$ and $y \in B$. Then the *domain* of the function f is defined to be the set of $x \in A$ for which there exists a $y \in B$ such that $(x,y) \in f$. The *range* of the function f is defined to be the set of $y \in B$ for which there exists an $x \in A$ such that $(x,y) \in f$. Let us return to our first example in which case we have $f = \{(1,b),(2,c),(3,a),(4,e),(5,e)\}$. The domain of f is given by the set $\{1,2,3,4,5\}$, and the range is given by the set $\{a,b,c,e\}$. Note in this case that the range is not equal to B. Notationally, we sometimes write $f : A \rightarrow B$ to mean the function maps the set A into the set B. In Fourier analysis we often work with functions that are described by an analytical expression or mathematical formula. For example, in

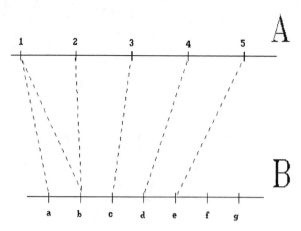

Figure 2.4 Graphical representation of a relation that is not a function.

Chapter 1 we computed the Fourier transform of the function given by the formula $f(x) = e^{-ax}$. We now relate this description to our formal definition of a function as a set of ordered pairs. In this case we let $A = B = R^1$ (the set of all real numbers). Then the function f can be considered as the infinite set of ordered pairs $\{x, f(x)\} = \{x, e^{-ax}\}$.

Recalling our definition of a Cartesian product we offer the following formal definition of a function: A *function* f is the Cartesian product of the sets A and B such that if $(x, y) \in f$ and $(x, z) \in f$ then $y = z$.

As specific examples, let's consider the relation given by the formula $f(x) = x^2$ or in Cartesian product form $\{(x, x^2)\}$. This relation is a function because every element x maps to a unique element x^2. For example, the element $x = 2$ maps only to the element $y = 4$. On the other hand, the relation given by the formula $f(x) = \sqrt{x}$ or $\{(x, \sqrt{x})\}$ is not a function because every element x maps to two separate elements $\sqrt[+]{x}$ and $\sqrt[-]{x}$. For example, $(4, 2) \in f$ and $(4, -2) \in f$, and clearly $2 \neq -2$.

We say a function $f : A \to B$ is *one-to-one* on A if, and only if, for every $x \in A$ and $y \in A$ we have $f(x) = f(y)$ implies $x = y$. What this means is that for every $z \in B$, there is only one $x \in A$ such that $z = f(x)$ or $(x, z) \in f$. For example, the function $f(x) = x^2$ is not one-to-one over the set of real numbers, since for any value $z = f(x) \in B$ we have both $+x$ and $-x$ mapping to z.

Two sets A and B are called *equivalent* if, and only if, there exists a one-to-one function f whose domain is the set A and whose range is the set B. That is to say, two sets are equivalent if there exists a one-to-one correspondence between their elements. Equivalent sets are said to have the same power.

We now consider a special function known as a *sequence*. A *finite sequence*, of n terms, is defined as a function whose domain is the set of integers $\{1, 2, 3, \ldots, n\}$. The range of the sequence is the set $\{f_1, f_2, \ldots, f_n\}$. The elements of the range are called *terms of the sequence*. In light of our formal definition, a sequence is given by the set of ordered pairs $\{(1,f_1),(2,f_2),...,(n,f_n)\}$. For example, consider the finite sequence given by the formula $f(n) = n^2$, $n = 1,2,3$. Then we have

$$f = \{(1,1),(2,4),(3,9)\}.$$

When we allow the domain of a sequence to include all the positive integers (i.e., let n go to ∞), then we have an *infinite sequence*.

If a set A is equivalent to the set of positive integers or equivalent to some (finite or infinite) subset of the positive integers, then A is said to be a *countable set*. In other words, if the elements of A can be placed in a one-to-one correspondence with the set of positive integers, then we call the set A countable. This can be thought of as saying the elements of A are the terms of a one-to-one sequence. For example, the infinite set whose terms are given as $f_n = n^2$ is a countable set because each element can be placed in a one-to-one correspondence with the positive integers. In other words, each term n^2 can be directly related to the positive integer n in a one-to-one correspondence.

DISTANCE AND METRIC SPACES

A *metric space* is simply a set for which we can talk about the distance between any two elements. For example, the set of real numbers can be considered a metric space if we define distance as the absolute value of the difference between any two elements. The problem now arises as how to make a metric space out of sets which consist of more complicated elements such as continuous functions or matrices. The key factor is to be able to impose a distance function on these sets. However, we must first clearly define what we mean by distance. Let A be any set, and to every pair of elements x,y of A, let us associate a real number (positive or zero). This real number (denoted as $d(z,y)$) is called a *distance function* if the following conditions are satisfied:

(1) $d(x,y) = 0$ if, and only if, $x = y$.
(2) $d(x,y) = d(y,x)$.
(3) $d(x,y) \leq d(x,z) + d(z,y)$.

While the above three rules are obvious when we think in terms of real numbers and define $d(x,y) = |x - y|$, they are actually much more general and powerful than this simple illustration of distance.

Distance on a set is important because it allows us to consider such concepts as convergence and continuity. In other words, if, for some space (or set), we can find a function $d(x,y)$ (regardless of how abstract) that satisfies the conditions of a distance function, then we can consider convergence and continuity.

It turns out that many of the sets of interest to us fall into the category of vector spaces over the real field. For many of these vector spaces it is possible to define a norm that, in turn, satisfies the conditions of a distance function. Let us consider a vector space $\mathbf{V}$ over the real field R^1. A *norm*, denoted as $m(x)$, is a mapping from $\mathbf{V}$ into the set of non-negative real numbers such that:

(1) $m(x) = 0$ if, and only if, $x = 0$.
(2) $m(x + y) \leq m(x) + m(y)$.
(3) $m(ax) = |a|m(x)$, where $x \in V$ and $a \in R^1$.

Notationally, we sometimes write $m(x) = |x|$. As we have already indicated, the set of all complex numbers C forms a vector space over the real field. We define the norm of a complex number $z = (x, y)$ as the square root of the sum of the squares of the real and imaginary portion of z, that is,

$$m(z) = \|z\| = (x^2 + y^2)^{1/2}.$$

We now proceed to show that this definition does indeed satisfy all of the above conditions for a norm.

(1) $m(z) = (x^2 + y^2)^{1/2} = 0$. This statement is true if, and only if, $x = y = 0$, or $z = 0$.
(2) $m(z_1 + z_2) \leq m(x_1, y_1) + m(x_2, y_2)$. To demonstrate this we must show

$$[(x_1 + x_2)^2 + (y_1 + y_2)^2]^{1/2} \leq [x_1^2 + y_1^2]^{1/2} + [x_2^2 + y_2^2]^{1/2}.$$

Squaring both sides, we determine that this will be true if

$$[x_1 + x_2]^2 + [y_1 + y_2]^2 \leq x_1^2 + y_1^2 + x_2^2 + y_2^2 + 2[(x_1^2 + y_1^2)(x_2^2 + y_2^2)]^{1/2}$$

or

$$x_1 x_2 + y_1 y_2 \leq [(x_1^2 + y_1^2)(x_2^2 + y_2^2)]^{1/2}$$

or (if again squaring both sides)

$$x_1^2 x_2^2 + y_1^2 y_2^2 + 2x_1 x_2 y_1 y_2 \leq x_1^2 x_2^2 + y_1^2 y_2^2 + x_1^2 y_2^2 + x_2^2 y_1^2$$

or

$$2x_1 x_2 y_1 y_2 \leq x_1^2 y_2^2 + x_2^2 y_1^2,$$

which is equivalent to

$$(x_1 y_2 - x_1 y_1)^2 \geq 0,$$

which is indeed true.
(3) For any $a \in R^1$ we have

$$m(az) = (a^2 x^2 + a^2 y^2)^{1/2} = (a^2)^{1/2}(x^2 + y^2)^{1/2} = |a|m(z).$$

If, for any vector space over the real field, we can find a function $m(x)$ which satisfies the prescribed conditions, then we have found a norm for that vector space. We now show that a norm satisfies all the conditions of a distance function. If we define the distance function $d(x,y)$ (on a vector space) as being equal to a norm of $x - y$ on that vector space, then we have:

(1) $d(x,y) = m(x - y) = 0$ if, and only if, $x - y = 0$ or $x = y$.

(2) $d(x,y) = m(x - y) = m(y - x) = d(y,x)$.

(3) $d(x,y) = m(x - y) = m(x - z + z - y) \leq m(x - z) + m(z - y)$
    $d(x,y) \leq d(x,z) + d(z,y)$.

We note that a norm is usually easier to work with than a distance function because it is defined in terms of a single element rather than two elements.

CONCEPT OF ORDER

The concept of order is quite familiar when we think in terms of the real number system. That is, for any two real numbers x and y, we must have either $x \leq y$ or $y \leq x$. This concept can be applied to many other sets if we first clearly define its properties. We begin with a relation denoted by the symbol T. The relation T is called an *ordered relation* if it satisfies the following conditions:

(1) It must be reflexive; that is, xTx must always be true.

(2) It must be antisymmetric; that is, if xTy and yTx, then $x = y$.

(3) It must be transitive; that is, if xTy and yTz, then xTz.

For the real number system the symbol $\leq$ is usually used instead of T. In this case, given any real numbers x, y, and z the above conditions become:

(1) It must be reflexive; that is, $x \leq x$ must always be true.

(2) It must be antisymmetric; that is, if $x \leq y$ and $y \leq x$, then $x = y$.

(3) It must be transitive; that is, if $x \leq y$ and $y \leq z$, then $x \leq z$.

For the real numbers, if $x \leq y$, then notationally we also write $y \geq x$. It is important to think of the reflexive, antisymmetric, and transitive properties as being crucial to the concept of an ordered relation, and we should attempt to think a little more generally than just the real number system.

As an example of a somewhat more complex ordering relation let us consider the set A that consists of all continuous functions that map R^1 to R^1. For any two elements $f,g \in A$ we must determine an order. We proceed as follows: For any particular value x we let $y_1 = f(x)$ and $y_2 = g(x)$, where y_1 and y_2 are real numbers. Therefore, for each particular value of x, we will have $y_1 \leq y_2$ or $y_2 \leq y_1$. For the sake of discussion suppose for $x = x_1$ we have $y_1 \leq y_2$, then we say $f \leq g$. However, for some other value of x, say x_2, we

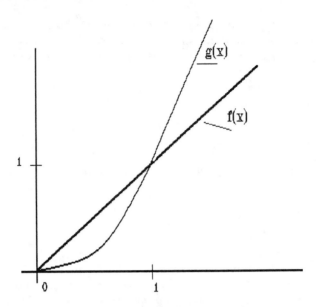

Figure 2.5   Ordering of continuous functions.

may have $y_2 = g(x_2) \le y_1 = f(x_2)$. Thus, we see that the concept of order for this set is a pointwise property. More specifically, if $f(x) = x$ and $g(x) = x^2$ and we consider the functions over the domain $[0, 2]$, we have:

$$g \le f \qquad \text{if} \quad x \in [0, 1],$$
$$f \le g \qquad \text{if} \quad x \in [1, 2].$$

This is shown graphically in Figure 2.5.

It is interesting to note that if we choose a different way to characterize the magnitude of a continuous function, we will obtain different results. For example, let

$$y_1 = \int_a^b f(x)\,dx, \quad \text{and} \quad y_2 = \int_a^b g(x)\,dx,$$

where the interval $[a, b]$ is the domain over which both functions are defined. Applying these results to our previous example, in which $f(x) = x$ and $g(x) = x^2$, we obtain

$$y_1 = \int_0^2 x\,dx = 2 \quad \text{and} \quad y_2 = \int_0^2 x^2\,dx = 8/3.$$

Thus we see that $y_1 \le y_2$, and this result is not dependent upon any single point in the interval. In other words, we defined the real numbers y_1 and y_2 in terms of the area under the individual functions over the entire domain. A set A is called *totally ordered* if, for *any* two elements $x, y \in A$, either $x \le y$ or $y \le x$ is satisfied. Obviously, the set of real numbers is totally ordered.

Depending upon the way in which we define our magnitude for a continuous function, we may consider the set of all continuous positive functions to be totally ordered. In other words, by our first choice of the values y_1 and y_2 being the value of the function at an individual point, we found the set was not totally ordered. On the other hand, when we assigned values to y_1 and y_2 based on the area under the curve of the continuous functions, we found that the set could be considered totally ordered.

CONCEPTS FROM TOPOLOGY

The basic idea behind topology is the study of limits or, equivalently, neighborhoods. The problem of interest to us is to determine if a sequence of elements from a set A converges to a fixed element of that set. Crucial to the question of convergence is the concept of distance, because when we say that a sequence of elements $x_n \in A$ converges to some fixed element $x \in A$, we require that the distance between each x_n and x (for sufficiently large n) be arbitrarily small. Therefore, before we are able to speak of convergence of a sequence we must be able to define a distance function $d(x,y)$ on the set. The most general approach to the problem of convergence is via the concept of neighborhoods. It turns out that if A has a distance function defined on it, we can define a neighborhood on the set A.

For any real number r, and any element x_0 of A, we define an *(open) neighborhood* of radius r about x_0 (denoted as $N_r(x_0)$) as the set of all $x \in A$ such that $d(x, x_0) < r$.

A *closed neighborhood*, $N_r(x_0)$, is defined as above, with the exception that we require $d(x, x_0) \le r$.

A *deleted neighborhood*, $N_r'(x_0)$, is an open neighborhood with the element x_0 (or center) removed.

In the set of real numbers, with distance defined as $d(x,y) = |x - y|$, we can see that a neighborhood of radius r about x_0 is simply the open interval $(x_0 - r, x_0 + r)$, or, mathematically, the set of all values of x satisfying the following inequality:

$$x_0 - r < x < x_0 + r.$$

A closed neighborhood of radius r about x_0 is given by the closed interval $[x_0 - r, x_0 + r]$, or, mathematically, the set of values of x satisfying

$$x_0 - r \le x \le x_0 + r.$$

Finally, a deleted neighborhood or radius r about the point x_0 is given by the union of the following two open intervals

$$(x_0 - r, x_0) \cup (x_0, x_0 + r).$$

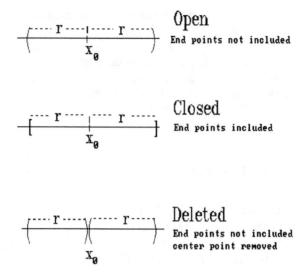

Figure 2.6   Neighborhoods in R^1.

These one-dimensional neighborhoods, or intervals, are graphically illustrated in Figure 2.6.

In the field of complex numbers using the norm $m(z) = (x^2 + y^2)^{1/2}$ as the distance function, we can see that a neighborhood of radius r about (x_0, y_0) is the interior of a circle (of radius r) centered at the point (x_0, y_0). A closed neighborhood would include the boundary of the circle, and a deleted neighborhood would exclude the center of the circle. These two-dimensional neighborhoods are graphically illustrated in Figure 2.7.

An element (or point) x_0 of a set A is called an *accumulation point*, (or limit point) if every neighborhood centered at x_0 contains an infinite number of points.

In the set of real numbers, every element is an accumulation point. This is because, given any $\varepsilon > 0$ and real number r_0, we have $r_n = (r_0 + \varepsilon/n) \in N_\varepsilon(r_0)$, for all $n > 0$. In other words, we can always find an infinite number of real numbers (r_n) that fall inside the neighborhood $N_\varepsilon(r_0)$.

As a counterexample, the set of integers does not have any accumulation points because, given any integer K, if we choose a neighborhood of radius $\frac{1}{4}$ about this integer K, then this neighborhood contains only the single element K.

LIMITS AND CONTINUITY

We are now in a position to discuss limits. There are two types of limits which are of interest to us, namely, the limit of a function and the limit of a sequence. We begin by considering limits of a function.

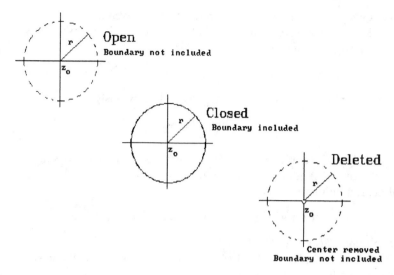

Figure 2.7   Neighborhoods in R^2.

Let $f : A \to B$ be a function with domain $S \subset A$ and range $T \subset B$. If a is an accumulation point of S and if $b \in B$, then when we write

$$\lim_{x \to a} f(x) = b$$

we mean the following: Given any $\varepsilon > 0$ there exists another real number $\delta > 0$ such that $x \in N'_\delta(a) \cap S$ implies $f(x) \in N_\varepsilon(b)$.

Note. We require $x \in N'_\delta(a) \cap S$, rather than $x \in N'_\delta(a)$, to make certain that x is in the domain of f. Also, we require the point a to be an accumulation point to guarantee that $N'_\delta(a) \cap S$ will never be an empty set.

In the above definition we do not require that $f(a) = b$ or that f even be defined at b. In the special case where (1) f is defined at a and $f(a) = b$ and (2) $\lim_{x \to a} f(x) = b = f(a)$, we say the function f is *continuous at a.*

We now look at a simple example to illustrate some of the principles involved in the above definitions. Consider the function $f : R^1 \to R^1$ whose domain S is the closed interval $[0,1]$ (Recall that this means f is defined for $0 \le x \le 1$). Furthermore, let this function be given by the rule

$$f(x) = 2x, \qquad 0 \le x \le 1,$$
$$f(1) = 0.$$

Let's consider

$$\lim_{x \to a} f(x) = b \qquad \text{for any} \quad a \in S.$$

Following the definition of the limit, we have: Given any $\varepsilon > 0$, if we choose $\delta = \varepsilon/2$ we can always guarantee that (for all $a \in S$) $x \in N_\delta'(a) \cap S$ implies $f(x) \in N_\varepsilon(b)$; in other words,

$$x \in ((a - \delta, a) \cup (a, a + \delta)) \cap S \quad \text{implies} \quad f(x) \in (b - \varepsilon, b + \varepsilon).$$

Let's first consider the limit as x goes to $\frac{1}{2}$. Obviously

$$\lim_{x \to 1/2} f(x) = 1.$$

f is clearly defined at $x = \frac{1}{2}$ and $f(\frac{1}{2}) = 1$. Thus, f is continuous at $x = \frac{1}{2}$. For the sake of discussion, suppose we are given $\varepsilon = .1$, then a choice of $\delta = .05$ will guarantee that if

$$x \in (.45, .5) \cup (.5, .55), \quad \text{then} \quad f(x) \in (.9, 1.1).$$

For example, $f(.46) = .92 \in (.9, 1.1)$.

Now let's consider the limit of f at another point in S, namely, $x = 1$. In this case we have

$$\lim_{x \to +1} f(x) = 2.$$

However, we note that while f is indeed defined at $x = 1$, the value is not equal to 2 ($f(1) = 0 \neq 2$). Thus, by our definition of continuity, f is not continuous at $x = 1$. Again, for the sake of discussion, suppose we are given $\varepsilon = .1$, then a choice of $\delta = .05$ will guarantee that if

$$x \in (.95, 1) \quad \text{then} \quad f(x) \in (1.9, 2.1).$$

Note that in this case if we only required $x \in N_\delta'(a)$ we would have

$$x \in (.95, 1) \cup (1, 1.05),$$

but f is not defined on the interval $(1, 1.05)$. Thus, we see the reason for requiring $x \in N_\delta'(a) \cap S$.

Another fact that this example illustrates is that the continuity of a function is a local property. That is to say, continuity is defined at a point of the domain. If a function is continuous over the entire domain S, then we say f is *continuous on* S. In the previous example, given any ε we were able to use the same δ (namely, $\delta = \varepsilon/2$) for all values of x in S. This, in general, will not be the case, and, in fact, δ will usually depend upon the element x. However, when the function f is continuous on S and the choice of δ does not depend upon x, we say that the function is *uniformly continuous* on the set S.

When we discuss sequences we have the following definition of a limit. If $\{x_n\}$ is a sequence of elements, from the set A, then when we write

$$\lim_{n \to \infty} x_n = a$$

we mean, given any $\varepsilon > 0$, there is a value N such that whenever $n > N$ we have $x_n \in N_\varepsilon(a)$.

This definition is really quite similar to that previously presented for the limit of a function. We require some condition on the elements of the domain (namely, $n > N$) which will guarantee that the terms of the sequence will be within an arbitrarily small neighborhood of the limit value a. We consider the condition $n > N$ to mean that n is within some deleted neighborhood of infinity.

As an example, consider the sequence defined as $x_n = 1/n^2$, which obviously converges to 0. Again, following the definition of the limit of a sequence, we find, given any $\varepsilon > 0$, if we choose $N = (1/\varepsilon)^{1/2}$ then $x_n \in N_\varepsilon(0)$ (or $|1/n^2| < \varepsilon$) for all values of n such that $n > N$. For example, let ε be given by .01; then $N = (1/.01)^{1/2} = 10$, and we find

$$\left|\frac{1}{n}\right|^2 < .01 \qquad \text{for all} \quad n > 10.$$

CONVERGENCE OF SEQUENCES

A sequence $\{x_n\}$ of elements of A is called *convergent* in A if there exists a point $a \in A$ such that

$$\lim_{n \to \infty} x_n = a.$$

In this case we say the sequence converges to a. A sequence that does not converge is said to *diverge*.

A sequence of elements $\{x_n\}$ from a set A is called a *monotonic increasing sequence* if $x_{n+1} \geq x_n$ for all n and, similarly, is called a *monotonic decreasing sequence* if $x_{n+1} \leq x_n$ for all n. For example, the sequence whose terms are given as $a_n = n^2$ is monotonically increasing. On the other hand, a sequence whose terms are given as $a_n = 1/n$ is monotonically decreasing.

A sequence $\{x_n\}$ is called *bounded* if there exists a real positive number B and a positive integer N such that $d(x_n, 0) \leq B$ for all $n > N$. That is to say, all but a finite number of terms lie within a neighborhood of radius B, about the origin.

Based on the above definition of bounded we have: A monotonic sequence (increasing or decreasing) is convergent if, and only if, it is bounded. While we will not prove this statement here, it is rather obvious that if a monotonic sequence is bounded, then the terms must start "bunching up" somewhere within a finite neighborhood of the origin. Similarly, if a sequence converges to a finite limit (accumulation) point, then it is always possible to construct a finite neighborhood about the origin which will include all but a finite number of terms of the sequence.

We have been quite general thus far when talking about convergence of a sequence; in other words, we have not really specified the types of sequences involved. Of particular interest to us will be sequences whose terms are functions. In this case we have the following concept known as *simple convergence* or *pointwise convergence*.

For each value of n let f_n be a function, with domain S, which maps the set A to the set B. For each $x \in S$ we define a sequence in B whose terms are given by $y_n = f_n(x)$. Now, for each x, we can consider convergence of the sequence $\{y_n\}$. More formally, given any $x \in S$ and $\varepsilon > 0$, we say $f_n(x)$ converges pointwise to $y = f(x)$, if there exists an integer N (depending on both ε and x) such that

$$n > N \quad \text{implies} \quad d(y_n, y) < \varepsilon.$$

For example, let us consider the sequence of functions given by $f_n(x) = x/n$. First we choose the point $x = 1$ to obtain the sequence of real numbers $y_n = 1/n$. This sequence obviously converges to 0. In other words, given any $\varepsilon > 0$ we can always find an N such that $|1/n| < \varepsilon$ whenever $n > N$. If we now choose the point $x = 2$, our sequence of real numbers turns out to be $y_n = 2/n$, which also converges to zero. However, in this case, for the same value of ε a larger value of N is required to guarantee that $2/n < \varepsilon$ for all $n > N$.

This type of pointwise convergence, while interesting, is not really strong enough for our needs. For example, it is possible to have each function $f_n(x)$ continuous and pointwise convergent (for all x) to a limit function $f(x)$ which is not itself continuous. What we desire is a convergence that eliminates this particularly undesirable property. Thus, we present the following stronger definition known as *uniform convergence*. We say that the sequence of functions $f_n(x)$ converges uniformly to the limit function $f(x)$ if, given any $\varepsilon > 0$, there exists an integer N such that

$$n > N \quad \text{implies} \quad d(f_n(x), f(x)) < \varepsilon \qquad \text{for all} \quad x \in S.$$

We note that the difference between pointwise convergence and uniform convergence is the fact that N does not depend on the value of x for uniform convergence. Examples of this type of convergence are presented in the next section.

The most useful fact about uniform convergence is that it passes continuity of the individual sequence functions over to the limit function. That is to say, if a sequence of continuous functions, $\{f_n(x)\}$, converges uniformly, then the limit function $f(x)$ is also continuous.

CHOICE OF A NORM FOR CONVERGENCE

Thus far, we have really said nothing about the distance function that we should use for a sequence of functions. The type of norm obviously depends upon the type of sequence functions. Let us now consider a few examples of a distance function, and, for the sake of illustration, let us assume that our functions map R^1 into R^1 over a finite domain $[a, b]$. We first recall, from our previous remarks, that if for a vector space we can define a norm, then we can define a distance function in terms of this norm. Thus, we concentrate our efforts on defining a norm. Perhaps the most familiar choice of a norm for this type of a sequence function is the absolute value of $f(x)$ at a particular location x within the interval $[a,b]$, that is,

$$m(f) = |f(x)| \quad \text{for each} \quad x \in [a,b].$$

Next, as another choice of a norm, let $m(f)$ be the maximum value of the absolute value of $f(x)$ over the interval $[a,b]$, that is,

$$m(f) = \max\{|f(x)| \text{ for all } x \in [a,b]\}.$$

Note. This definition of the norm does not depend upon any particular value of x in the interval. Thus, if we have convergence in terms of this norm, it is necessarily uniform convergence.

As a final example of a norm for this type of function we choose

$$m(f) = \int_a^b |f(x)| dx.$$

This norm is also independent of any particular value of x within the interval $[a,b]$. Let's be even more specific in an attempt to clarify some of the previous remarks. Consider the sequence of functions given by

$$f_n(x) = x^2/n \quad \text{for} \quad x \in [-1,1].$$

We can easily see that

$$\lim_{n \to \infty} f_n(x) = 0 \quad \text{for all} \quad x \in [-1,1].$$

Now let's consider convergence in terms of our first definition of a norm, that is,

$$m(f_n) = |f_n(x)| = \left|\frac{x^2}{n}\right| = \frac{x^2}{n}.$$

Thus,

$$d(f_n,f) = d(f_n,0) = m(f_n) = \frac{x^2}{n}.$$

Now for all values of $x \in [-1,1]$, we have: Given any $\varepsilon > 0$, if we choose N to be an integer greater than x^2/ε, then

$$n > N \quad \text{implies} \quad d(f_n,0) < \varepsilon.$$

For example, if $x = .5$ and $\varepsilon = .01$, we determine $N = 25$. We see by this definition of the norm that N depends upon both x and ε. However, in this situation we can find a worst-case value for N. Therefore, for all x within the interval $[-1,1]$, given any $\varepsilon > 0$, if we choose N to be an integer greater than $1/\varepsilon$ (since $|x| \le 1$), we will satisfy the conditions of convergence and, thus, the sequence is uniformly convergent to zero.

Now let us use the norm defined as

$$\int_{-1}^{1} |f_n(x)|\, dx = \int_{-1}^{1} \frac{x^2}{n}\, dx = \frac{2}{3n}.$$

Therefore,

$$d(f_n, f) = d(f_n, 0) = m(f_n) = \frac{2}{3n}.$$

By this definition of the norm we have: Given any $\varepsilon > 0$, if we choose N to be an integer greater than $2/3\varepsilon$, then $d(f_n, 0) < \varepsilon$. Thus, we have established uniform convergence.

We now present a somewhat more stimulating example. Let $f_n(x) = e^{-nx}$ be defined on the semi-open interval $(0, 1]$. Again we see

$$\lim_{n \to \infty} f_n(x) = 0 \qquad \text{for all} \quad x \in (0, 1].$$

Now using the absolute value norm we have

$$d(f_n, 0) = m(f_n) = |e^{-nx}| = e^{-nx}.$$

Again using the definition, for any $x \in (0, 1]$ we have: Given any $\varepsilon > 0$, if we choose N to be an integer greater than

$$\left(\frac{1}{x} \right) \ln \left(\frac{1}{\varepsilon} \right),$$

then

$$n > N \quad \text{implies} \quad m(f_n(x)) < \varepsilon.$$

Recall that in the previous example we were able to remove the dependence of N on x by choosing a worst-case situation. However, here it is not possible. Even though x never equals 0 it can become arbitrarily close to 0 and, therefore, $1/x$ can become arbitrarily large. Thus, using this norm we can see that the sequence is pointwise convergent but not uniformly convergent. Now let's consider what happens if we use as the norm

$$m(f_n) = \int_{0+}^{1} |f_n(x)|\, dx = \int_{0+}^{1} e^{-nx}\, dx = \frac{1 - e^{-n}}{n},$$

which implies

$$d(f_n, f) = d(f_n, 0) = m(f) = \frac{1 - e^{-n}}{n}.$$

Thus, given any $\varepsilon > 0$, if we choose N such that

$$\frac{1 - e^{-N}}{N} < \left(\frac{1}{N} \right) < \varepsilon$$

or

$$N > \frac{1}{\varepsilon},$$

then we have

$$d(f_n, 0) < \varepsilon.$$

In this case we see that N is independent of the value of x within the interval $(0,1]$ and, therefore, by this definition of the norm we have that f_n is uniformly convergent.

CAUCHY CONVERGENCE

The type of convergence that we have been discussing up to this point assumes that a sequence converges to some definite fixed element within the set. We required, for sufficiently large n, that the distance between the terms of the sequence and this limit point be made arbitrarily small. We now consider another type of convergence (known as *Cauchy convergence*) in which we require, for sufficiently large n, that the terms of the sequence become arbitrarily close to each other.

Mathematically, we say a sequence of terms $\{x_n\}$ is a *Cauchy convergent* if, given any $\varepsilon > 0$, there exists an integer N such that $n > N$ and $m > N$ implies $d(x_n, x_m) < \varepsilon$.

We can easily show that every sequence $\{x_n\}$ from the set A which is convergent to a limit point $a \in A$ is also a Cauchy sequence. To do this we assume

$$\lim_{n \to \infty} x_n = a;$$

thus, given any $\varepsilon/2 > 0$, there exists an integer N such that

$$n > N \quad \text{implies} \quad d(x_n, a) < \frac{\varepsilon}{2}$$

and

$$m > N \quad \text{implies} \quad d(x_m, a) < \frac{\varepsilon}{2}.$$

Therefore, by the properties of the distance function, we have

$$d(x_n, x_m) \leq d(x_n, a) + d(x_m, a) < \frac{\varepsilon}{2} + \frac{\varepsilon}{2} = \varepsilon.$$

Thus we have shown, given any $\varepsilon > 0$, there exists an N such that $n > N$; $m > N$ implies $d(x_n, x_m) < \varepsilon$ and thus, by definition, the sequence $\{x_n\}$ is Cauchy convergent.

The converse of the previous statement, however, is not true in general. That is to say, if a sequence of terms $\{x_n\}$ from a set A is Cauchy convergent then the sequence will indeed converge to some element a, but we can not say for sure that this element will be in the set A. For example, consider the sequence of rational numbers given by

$$r_n = \left(1 + \frac{1}{n}\right)^n.$$

This sequence is a Cauchy sequence that converges to the number $e = 2.718281828...$, which is not a rational number.

A set in which every Cauchy sequence converges to an element of the set is called a *complete set*. For example, the set of real numbers is a complete set because every sequence of real numbers converges to a real number.

When we deal with Cauchy convergence of a sequence of functions we again have pointwise and uniform convergence. Analogous to our previous considerations of convergence of sequences we have: If N is dependent upon x then we have simple (or pointwise) Cauchy convergence, while if N is independent of x then the sequence is called *uniformly Cauchy convergent*.

SUMMARY

In this chapter we presented basic mathematical, or analytical, terminology and concepts. We began by discussing sets in general and then limited our attention to several useful sets such as fields and vector spaces. A proper definition of a function (and a sequence) as a mapping from one set to another was presented. We showed that when a distance function could be defined on a set we were able to consider convergence of functions and sequences. The material presented in this chapter is used as a foundation upon which the remaining chapters can build and develop other topics.

PROBLEMS

1   Given the following sets

$$A = \{1,2,3,4,9,10,11,12\},$$
$$B = \{1,2,3,4\},$$
$$C = \{9,10,11,12\},$$
$$D = \{5,6,7,8\},$$

describe the following sets

(a) $A \cup D$ (f) $(C \cap A) \cup B$
(b) $A \cap D$ (g) $C \cap (A \cup B)$
(c) $B - A$ (h) $(A - B) \cap C$
(d) $A - B$ (i) $(B - A) \cap C$
(e) $(A \cup B) \cap C$ (j) $(D - A) \cup C$

2   Given any two complex numbers $z_1 = (x_1, y_1)$ and $z_2 = (x_2, y_2)$, addition is defined as $z_1 + z_2 = (x_1 + x_2, y_1 + y_2)$. Multiplication is defined as

$z_1 z_2 = (x_1 x_2 - y_1 y_2, x_1 y_2 + x_2 y_1)$. Show that this sum law and product law are both associative and commutative.

3 For the field of complex numbers show that the inverse element is indeed given as $z^{-1} = \left(\dfrac{x}{x^2 + y^2}, \dfrac{-y}{x^2 + y^2} \right)$; that is, show that $z z^{-1} = z^{-1} z = (1, 0)$.

4 Do the following sets satisfy the conditions required to be considered a group under the properly defined addition law (why or why not)?

    (a) set of real numbers,
    (b) set of complex numbers,
    (c) set of all integers,
    (d) set of all positive integers,
    (e) set of all positive real numbers.

5 Do the following sets satisfy the conditions required to be considered a field under the properly defined addition and multiplication laws (why or why not)?

    (a) set of real numbers,
    (b) set of complex numbers,
    (c) set of all integers,
    (d) set of all positive integers,
    (e) set of all positive real numbers.

6 Consider the definition of a field. If we remove condition (d) and do not require an inverse under multiplication, then we have the special set known as a *ring*. Show that the set of integers forms a ring.

7 Show that the complex number system forms a vector space over the real field.

8 Show that the set of all polynomials in x, with real coefficients, forms a vector space over the field or real numbers.

9 Show that the set of all integrable real functions forms a vector space over the real field.

10 Does a one-to-one function mean that the function is necessarily linear (*Hint*: Show result by counter example).

11 Consider the vector space A over the real field to be the set of all integrable real positive functions on the interval $[0, 1]$. For any function $f \in A$ and $a \in R^1$, show that

$$m(f) = \int_0^1 f(x)\, dx$$

satisfies the definition of a norm for the set A.

12 In this chapter we showed that for the real number system, $\leq$ (less than or equal) is an ordered relation.

(**a**) Why isn't $<$ (less than) an ordered relation?
(**b**) Is $\geq$ (greater than or equal to) an ordered relation?

13 Give a mathematical and physical description of (a) an open, (b) a closed, and (c) a deleted neighborhood in R^3.

14 Suppose that for some metric space we know

$$\lim_{n\to\infty} a_n = a \quad \text{and} \quad \lim_{n\to\infty} b_n = b.$$

(**a**) Show that
$$\lim_{n\to\infty} (a_n + b_n) = a + b.$$

(**b**) Show that
$$\lim_{n\to\infty} d(a_n, b_n) = d(a,b),$$

where d is the distance function for that metric space.

15 Let f and g be real-valued functions of a metric space A, with a distance function d. Let p be an accumulation point of A and assume

$$\lim_{x\to p} f(x) = a,$$

$$\lim_{x\to p} (f(x) + g(x)) = a + b,$$

$$\lim_{x\to p} (f(x)g(x)) = ab.$$

16 Show that the sequence $\{a_n\}$ whose terms are given by $a_n = n^2$ is monotonically increasing.

17 Show that the sequence $\{a_n\}$ whose terms are given by $a_n = 1/n^2$ is monotonically decreasing.

BIBLIOGRAPHY

Apostal, T. M., *Mathematical Analysis*, Addison-Wesley, Reading, Mass., 1974.

Boas, R. P., *A Primer of Real Functions*, Carus Monograph No. 13, John Wiley & Sons, New York, 1960.

Hasser, N. B., J. P. Lasalle, and J. A. Sullivan, *Intermediate Analysis*, Blaisdell, New York, 1964.

Rudin, W., *Principles of Mathematical Analysis*, McGraw-Hill, New York, 1964.

3

INTEGRATION THEORY

Both the Fourier series and Fourier transform require the evaluation of integral expressions. Consequently, the theory of integration plays a very important role in our study and understanding of Fourier analysis. In this chapter we walk through the basic theory of integration from a Lesbegue point of view. The material presented here is not intended to be a complete and/or thorough treatment of the advanced theory of Lebesgue integration. Instead, it is included to familiarize the reader with the basic concepts and ideas behind this approach to integration theory.

We develop the Lebesgue theory by introducing the step function and its integral. With this definition in hand we proceed to sequences of step functions and discuss their convergence. When a Cauchy sequence of step functions converge, we denote the limit function to which it converges as a *Lebesgue integrable function*. We further show that the sequence of real numbers formed by the integral of each term of the step function sequence also forms a converging Cauchy sequence.

As it turns out, this Lebesgue point of view allows us to conveniently study convergence of sequences and series of functions. The actual evaluation of a Lebesgue intetral is accomplished by relating it to the Riemann integral and then using the Fundamental Theorem of Calculus. These are the topics and points of view that will be considered in this chapter.

LEBESGUE MEASURE

We begin by limiting our attention to the space R^n and discuss the concept of Lebesgue measure. We first point out that R^n is the set of all ordered n-tuples

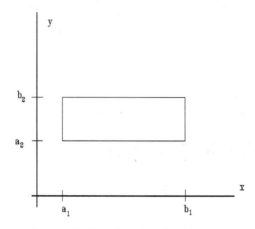

Figure 3.1 Two-dimensional open interval.

of real numbers. For example, R^1 is the set of all real numbers, R^2 is the set of all complex numbers, and so on. While most of our work will be with R^1 and R^2, we keep our definitions general at this point.

In the previous chapter we discussed an open interval in R^1. We now generalize the concept to an n-dimensional interval.

An *n-dimensional open interval* is a set of ordered n-tuples of the form

$$I = \{(x_1, x_2, \ldots, x_n) \text{ such that } a_k < x_k < b_k, \; k = 1, 2, \ldots, n\}.$$

Consider for example the two-dimensional open interval in R^2 given by

$$I = \{(x, y) \text{ such that } a_1 < x < b_1, \; a_2 < y < b_2\}.$$

This is shown in Figure 3.1.

Open n-dimensional intervals are very similar to open n-dimensional neighborhoods. Whereas a neighborhood may be thought of as an n-dimensional sphere, an interval may be thought of as an n-dimensional rectangle.

The length of an interval in R^1 is obviously $(b - a)$. Thus by analogy, for an interval in R^n we define the length as the product of the individual interval lengths, that is,

$$L(I) = (b_1 - a_1)(b_2 - a_2)\ldots(b_n - a_n).$$

In R^2 the length of an interval is the area of a rectangle and in $R^3 L(I)$ is the volume of a box.

We are now in a position to define a *Lebesgue covering*. Consider a set $S \subset R^n$; if there exists a countable collection of n-dimensional intervals,

$$I_L = \{I_1, I_2, \ldots\},$$

such that

$$S \subset \bigcup_{k=1}^{\infty} I_k,$$

Figure 3.2   Interval in R^1.

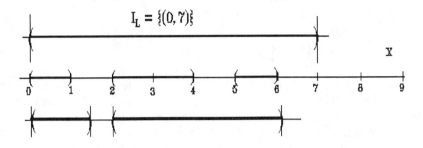

$$I_L = \{(0,1.5),(2,6.1)\}$$

Figure 3.3   Example of covering in R^1.

then I_L is called a *Lebesgue covering of S*.

The number

$$m(S) = \text{g.l.b.}\{L(I_k): \text{ where } \{I_k\} \text{ is a Lebesgue covering of } S\}$$

is properly called the *n-dimensional outer Lebesgue measure of S*. (Note that g.l.b. is the *greatest lower bound* on I_k for $k = 1$ to ∞). In our work we will simply call $m(S)$ the measure of S.

Let's look at some examples to illustrate the concepts. First consider $S \subset R^1$ to be the set consisting of the intervals $(0,1)$, $(2,4)$, and $(5,6)$ as illustrated in Figure 3.2.

There are many Lebesgue coverings of this set; for example, $I_L = \{(0,7)\}$ is one such covering with length $L(I_L) = 7$. $I_L = \{(0,1.5),(2,6.1)\}$ is another with length $L(I_L) = 5.6$. See Figure 3.3.

By definition, $m(S)$, the measure of S, is the greatest lower bound of the length of all possible coverings of S. Obviously, in this case, it is given by

$$m(S) = L(I_L) = 4, \quad \text{where } I_L = \{(0,1),(2,4),(5,6)\}.$$

In this example, the Lebesgue covering is simply the set itself. However, suppose the set S is now given by the closed interval $[0,1]$. Now we see the reason why we require the greatest lower bound and not just the smallest length. In this case, any open interval that contains $[0,1]$ (e.g., $(-1,2)$) will be a Lebesgue covering, but the greatest lower bound of the length of all such coverings is $m(S) = 1$. Note: We cannot have $(0,1)$ covering $[0,1]$ because the endpoints 0 and 1 are not elements of $(0,1)$.

We say a set S has *measure zero* if $m(S) = 0$. For example, in R^1 any set consisting of discrete points has measure zero (a point has no length). Similarly, in R^2 all lines have measure zero (again, a line in two dimensions has no area).

We say a property holds *almost everywhere* (written as a.e.) on $S \subset R^n$ if it is true for all $x \in S$ except on some subset of measure zero. For example, consider the function $f(x)$ defined over $[0,2]$ as follows:

$$f(x) = \begin{cases} x^2, & 0 \le x < 1, \quad 1 < x \le 2, \\ 0, & \text{if} \quad x = 1. \end{cases}$$

In this case we say $f(x) = x^2$ a.e. on $[0,2]$ since the only place where this is not true is the point $x = 1$, which is a subset of $[0,2]$ with measure zero. As another example, let $f(x) = x^2$ for all $x \in R^1$, except when x is a positive integer, in which case we set $f(x) = 0$. Again we have $f(x) = x^2$ a.e. on R^1 because the subset of all integers forms a set of measure zero in R^1.

STEP FUNCTIONS AND THEIR INTEGRALS

In this section our approach is to first define what is meant by a step function and the integral of a step function. We then consider special sequences of these step functions, which converge to other functions, known as *Lebesgue integrable functions*. Finally, the integrals of these Lebesgue functions are then defined in terms of (actually limits of) the integrals of the step function sequences.

We begin by considering a closed interval $[a,b]$ and a partition $P = \{x_0, x_1, \ldots, x_n\}$ (where $x_{k-1} < x_k$). We define a *step function* on the interval $[a,b]$ as

$$s(x) = c_k \quad \text{if} \quad x \in (x_{k-1}, x_k), \quad k = 1, 2, \ldots, n.$$

That is to say, $s(x)$ is constant on every *open* subinterval (x_{k-1}, x_k) of $[a,b]$. Note that $x_0 = a$ and $x_n = b$. An example of a step function is shown in Figure 3.4.

The integral of $s(x)$ over each subinterval $[x_{k-1}, x_k]$ is defined as $c_k(x_k - x_{k-1})$. Notationally, we write

$$\int_{x_{k-1}}^{x_k} s(x)\,dx = c_k(x_k - x_{k-1}).$$

From the above equation we easily generalize the definition to the integral of $s(x)$ over $[a,b]$ as follows:

$$\int_a^b s(x)\,dx = \sum_{k=1}^{n} c_k(x_k - x_{k-1}).$$

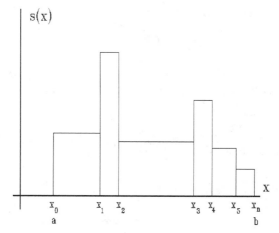

Figure 3.4 Example of step function.

A few comments are in order at this point. We first note that when we define a step function over $[a,b]$, we do not specify a value of $s(x)$ at the endpoints of the subintervals. In other words, $s(x)$ is defined as a constant over the interior of the open subinterval (x_{k-1}, x_k), but not necessarily specified at the endpoints. The integral of $s(x)$, on the other hand, is defined over each closed subinterval $[x_{k-1}, x_k]$, (i.e., it includes the endpoints). As can be seen from the definition, the integral is independent of the value of $s(x)$ at these endpoints. From our earlier remarks on the measure of a set, we see that the endpoints constitute a subset of $[a,b]$ of measure 0. Thus, we see that the integral of $s(x)$ over $[a,b]$ is independent of the behavior of $s(x)$ on a set of measure 0. This can be made intuitively more pleasing if we consider the integral of $s(x)$ over $[a,b]$ to be the sum of areas of the individual rectangles formed by each subinterval. In R^1 any set of measure 0 has no width and, thus, can contribute nothing to the total area.

Note. Our definition of

$$\int_a^b s(x)\,dx$$

is exactly equal to the Riemann integral of the step function $s(x)$ over $[a,b]$. However, by defining the integral as we have done, the Lebesgue theory can be developed without a previous knowledge of Riemann integration theory.

It is convenient to generalize the definition of a step function to intervals other than closed and connected ones. We do this in a very simple and straightforward way. Let I be any general interval (open, closed, half-open, disjointed, etc.) and let $[a,b]$ be a closed interval containing I (i.e., $[a,b] \supset I$). Now consider s to be a step function defined on $[a,b]$, such that $s(x) = 0$ for

$x \in ([a,b] - I)$. Then the integral of s over I, denoted as

$$\int_I s(x)\,dx$$

is simply defined to be the integral of s over $[a,b]$ as previously defined.

The following theorem gives some useful properties of the integrals of step functions.

Theorem 3.1 (Linearity and Order). If s and t are step functions defined over I, then we have

(a) $\displaystyle \int_I (s+t) = \int_I s + \int_I t,$

(b) $\displaystyle \int_I cs = c \int_I s,$ where c is any real number,

(c) $\displaystyle \int_I s \leq \int_I t,$ if $s(x) \leq t(x)$ for all $x \in I,$

(d) $\displaystyle \left| \int_I s \right| \leq \int_I |s|.$

While these results are rather obvious, we present the proof of part (a) to give an indication of the methods used to prove this type of theorem.

Proof. We must first clearly indicate what we mean by the addition of two step functions. Let $s(x)$ be a step function defined on the interval $[x_0, x_n]$ with a partition $P_1 = \{x_0, x_1, \ldots, x_n\}$ such that

$$s(x) = c_k \quad \text{for} \quad x \in (x_{k-1}, x_k), \quad k = 1, 2, \ldots, n.$$

Similarly, let $t(x)$ be a step function, defined on the interval $[y_0, y_m]$, with a partition $P_2 = \{y_0, y_1, \ldots, y_m\}$ such that

$$t(x) = d_k \quad \text{for} \quad x \in (y_{k-1}, y_k), \quad k = 1, 2, \ldots, m.$$

The problem with defining (and integrating) the sum of $s(x)$ and $t(x)$ is that both step functions are defined over different intervals with different partitions. Thus, we must obtain a common interval and partition.

Let $P = P_1 \cup P_2 = \{z_0, z_1, \ldots, z_N\}$, where

$$z_0 = \min(x_0, y_0),$$

$$z_N = \max(x_n, y_m),$$

and

$$z_{k-1} \leq z_k, \quad \text{for all} \quad k = 1, 2, \ldots, N.$$

We now redefine the step functions over this new interval $[z_0, z_N]$ as follows:

$$s(x) = c_k, \quad \text{for} \quad x \in (z_{k-1}, z_k), \quad k = 1, 2, \ldots, N,$$

where

$$\mathbf{c}_k = \begin{cases} c_k, & \text{if } x \in (x_{k-1}, x_k) \quad \text{for } x_{k-1}, x_k \in P_1, \\ 0, & \text{otherwise.} \end{cases}$$

A similar redefinition holds for $t(x)$ in terms of constants $\mathbf{d}_k$. Since we now have both functions defined over the same interval with the same partition, addition is simply given as

$$s(x) + t(x) = \mathbf{c}_k + \mathbf{d}_k \quad \text{if } x \in (z_{k-1}, z_k), \quad k = 1, 2, \ldots, N.$$

With a little thought the following statements become obvious:

$$\int_I s = \sum_{k=1}^{N} \mathbf{c}_k (z_k - z_{k-1}) = \sum_{k=1}^{n} c_k (x_k - x_{k-1}),$$

$$\int_I t = \sum_{k=1}^{N} \mathbf{d}_k (z_k - z_{k-1}) = \sum_{k=1}^{m} d_k (y_k - y_{k-1}).$$

Thus, we have

$$\int_I (s + t) = \sum_{k=1}^{N} (\mathbf{c}_k + \mathbf{d}_k)(z_k - z_{k-1})$$

$$= \sum_{k=1}^{N} \mathbf{c}_k (z_k - z_{k-1}) + \sum_{k=1}^{N} \mathbf{d}_k (z_k - z_{k-1})$$

$$= \sum_{k=1}^{n} c_k (x_k - x_{k-1}) + \sum_{k=1}^{m} d_k (y_k - y_{k-1}),$$

$$\int_I (s + t) = \int_I s + \int_I t. \qquad\qquad \text{Q.E.D.}$$

LEBESGUE FUNCTIONS

Recalling the definition of a vector space, it can be shown that the set of all step functions forms a vector space V_s over the real field. This vector space also has two additional properties: (1) If $s(x) \in V_s$, then $|s(x)| \in V_s$. (2) V_s is ordered; that is, for $s(x), t(x) \in V_s, s(x) \le t(x)$ has a meaning.

We define a norm on V_s as follows:

$$m(V_s) = \int_I |s| = \sum_{k=1}^{n} |c_k|(x_k - x_{k-1}).$$

It can easily be shown that this definition of $m(V_s)$ does indeed satisfy all required properties of a norm. Consequently, we are able to define a distance function and talk about convergence in this space.

We say a sequence of step functions $\{s_n\} \in V_s$ converges, almost everywhere on I (notationally: a.e. on I) to a function $s \in V_s$, if, given any $\varepsilon > 0$, there exists an integer N such that $n > N$ implies

$$\int_I |s_n - s| < \varepsilon.$$

A sequence of step functions $\{s_n\} \in V_s$ is a (uniformly convergent) Cauchy sequence, a.e. on I if, given any $\varepsilon > 0$, there exists an integer N such that $n > N$ and $m > N$ imply

$$\int_I |s_n - s_m| < \varepsilon.$$

We have previously referred to the fact that the value of the integral of a step function is independent of the behavior of the step function on a set of measure zero. Also, recall that when we say a step function converges a.e. on I, we mean that it converges for all $x \in I$ except on some subset of I which has measure zero. Therefore, by our choice of the norm we have somewhat sidestepped the ticklish question of almost everywhere properties. If, for example, we used as the norm the pointwise absolute value function

$$m(V_s) = |s_n(x)|,$$

then we would have to be concerned with those points (on a set of measure zero) for which the sequence $\{s_n(x)\}$ fails to converge.

In the previous chapter we showed (in the general situation) that a sequence which converges to some limit element in the set is a Cauchy sequence. We now go through this again for a sequence of step functions.

Theorem 3.2. If $\{s_n\} \in V_s$ is a sequence of step functions that converges, a.e. on I, to a limit function $s \in V_s$, then $\{s_n\}$ is a Cauchy sequence.

Proof. By assumption

$$\lim_{n \to \infty} s_n = s;$$

thus, given any $\varepsilon > 0$, there exists an integer N such that $n > N$ and $m > N$ imply

$$\int_I |s_n - s| < \frac{\varepsilon}{2}$$

and

$$\int_I |s_m - s| < \frac{\varepsilon}{2}.$$

Therefore,

$$\int_I |s_m - s_n| = \int_I |s_m - s + s - s_n|$$

$$\leq \int_I |s_m - s| + \int_I |s_n - s|$$

$$\leq \varepsilon/2 + \varepsilon/2 = \varepsilon. \qquad\qquad \text{Q.E.D.}$$

Just as in our previous work with rational numbers, the converse of this theorem is not true in general. That is to say, if a sequence of step functions is a Cauchy sequence, then it does indeed converge to some limit function, but that limit function is not necessarily a step function. In other words, V_s is not a complete space. For example, consider the step function sequence defined as

$$s_n(x) = c_k - \frac{1}{n} \qquad \text{for} \quad x \in (x_{k-1}, x_k).$$

Clearly, in the limit as n approaches infinity, this sequence of step function converges to the step function

$$s(x) = c_k \qquad \text{for} \quad x \in (x_{k-1}, x_k)$$

Now let us consider the sequence of step functions defined in the following way: Consider the interval $[a, b]$ to be partitioned into equal increments of width $\Delta x = (b - a)/n$, where n is allowed to range from 1 to infinity. Furthermore, let $x_k = x_{k-1} + \Delta x$. We define the step function sequence as

$$s_n(x) = c_k \qquad \text{for} \quad x \in (x_{k-1}, x_k) \qquad \text{where} \quad c_k = (x_{k-1})^2.$$

This step function is shown in Figure 3.5 for inceasing values of n. As can be clearly seen from this figure, as well as from the above equation, this sequence of step functions converges to $f(x) = x^2$. In other words,

$$\lim_{n \to \infty} s_n(x) = x^2$$

We now consider a rather interesting property of Cauchy sequences of step functions.

Theorem 3.3. If $\{s_n\}$ is a sequence of step functions which is Cauchy convergent, a.e. on I, then the sequence formed by the integrals of each term s_n is also a Cauchy sequence (based on the absolute value norm).

Proof. By assumption, $\{s_n\}$ is a Cauchy sequence. Therefore, given any $\varepsilon > 0$, there exists an integer N such that $n > N$ and $m > N$ imply

$$\int_I |s_m - s_n| < \varepsilon.$$

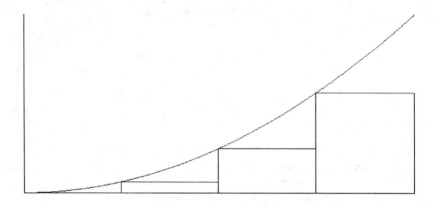

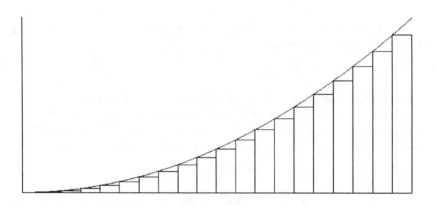

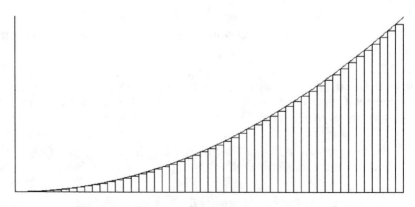

Figure 3.5 Sequence of step functions convering to x^2.

Thus,

$$\left| \int_I s_m - \int_I s_n \right| = \left| \int_I (s_m - s_n) \right| \le \int_I |s_m - s_n| < \varepsilon. \qquad \text{Q.E.D.}$$

What we have just shown is quite important. It states that the numbers equal to the integrals of s_n form a Cauchy sequence in the space of real numbers. In plain English, Theorem 3.3 tells us that if $\{s_n\}$ is a Cauchy sequence of step functions, then $\{s_n\}$ and $\{\int s_n\}$ converge. $\{s_n\}$ converges to a limit function f, and $\{\int s_n\}$ converges to a real number (recall that R^1 is a complete space and, thus, contains all its limit points). Notationally, we write

$$\lim_{n \to \infty} s_n = f$$

and

$$\lim_{n \to \infty} \int_I s_n = A.$$

The limit function f will not, in general, be a step function. However, when it is, we are able to show

$$A = \int_I f = \lim_{n \to \infty} \int_I s_n.$$

Since both $\{s_n\}$ and $\{\int s_n\}$ converge, given any $\varepsilon > 0$, there exists an integer N such that for all $n > N$ we have

$$\int_I |f - s_n| < \frac{\varepsilon}{2} \quad \text{and} \quad \left| \int_I s_n - A \right| < \frac{\varepsilon}{2}.$$

However, by previously demonstrated properties of step functions we have

$$\left| \int_I f - \int_I s_n \right| = \left| \int_I (f - s_n) \right| \le \int_I |f - s_n| < \frac{\varepsilon}{2}.$$

Therefore,

$$\left| \int_I f - A \right| = \left| \int_I f - \int_I s_n + \int_I s_n - A \right|$$

$$\le \left| \int_I f - \int_I s_n \right| + \left| \int_I s_n - A \right|$$

$$\le \frac{\varepsilon}{2} + \frac{\varepsilon}{2} = \varepsilon,$$

which implies

$$\int_I f = A.$$

Now let's consider the general case where the limit function is not a step function. When a sequence of step functions converges, a.e. on I, to a function f, we call f a *Lebesgue integrable function* (or simply a Lebesgue function) and denote it notationally as $f \in L(I)$. When f was a step function we were able to use the properties of the integral of step functions to show

$$\int_I f = \lim_{n \to \infty} \int_I s_n.$$

In the general case, we simply define the integral of f as this limit; that is, if

$$\lim_{n \to \infty} s_n = f, \qquad \text{a.e. on } I,$$

then, by definition (since we know $\int s_n$ converges), we have

$$\int_I f = \lim_{n \to \infty} \int_I s_n.$$

It is most important to note that this is the definition of the integral of the limit function f. In other words, we first determine the limit of the integrals of the sequence of step functions and then define this limit to be equal to the integral of f. This process is known as completing the space S_n. (*Note*: We sometimes say that s_n generates f.)

The next theorem shows that integrals of Lebesgue functions defined in this way satisfy several of the same properties as do integrals of step functions (as well as the same properties of the Riemann integral).

Theorem 3.4. Assume $f \in L(I)$ and $g \in L(I)$; then

(a) $(f + g) \in L(I)$ and $\int_I (f + g) = \int_I f + \int_I g.$

(b) $cf \in L(I)$ for every real constant c and

$$\int_I cf = c \int_I f.$$

(c) $\int_I f \leq \int_I g,$ if $f(x) \leq g(x)$ a.e. on I.

(d) $\left| \int_I f \right| \leq \int_I |f|.$

If we compare this to Theorem 3.1 we see that the integral of step functions and Lebesgue functions satisfy the same properties. It is important to note that the integral of a Lebesgue function is defined as the limit of the sequence of integrals of the generating sequence of step functions. Theorem 3.4 illustrates that many of the properties of the integrals of step functions are "passed over" to the integral of the limit function f. Again, we lightly go over the proof of part (a) to help illustrate the concepts. We first point out that when we define

$$\int_I f = \lim_{n \to \infty} \int_I s_n$$

we agree that it makes sense to write

$$\int_I |f - s_n|.$$

In other words, we define the norm of V_s to also be valid for $L(I)$. Now continuing with the proof of part (a): Suppose $\{s_n\}$ generates f and $\{t_n\}$ generates g; then, by definition, both s_n and t_n are Cauchy sequences and, thus, their sum $\{s_n + t_n\}$ is also a Cauchy sequence. Therefore, $\{s_n + t_n\}$ converges to a Lebesgue integrable function h, and $\{\int (s_n + t_n)\}$ converges to a real number h. We must first show that $h = f + g$, or, equivalently, $d(h,(f + g)) = 0$. Given any $\varepsilon > 0$, we can find an integer N such that $n > N$ implies each of the following:

$$\int_I |f - s_n| < \frac{\varepsilon}{3},$$

$$\int_I |g - t_n| < \frac{\varepsilon}{3},$$

$$\int_I |h - (t_n + s_n)| < \frac{\varepsilon}{3}.$$

Now,

$$d(h,(f + g)) = \int_I |h - (f + g)|$$

$$= \int_I |h - (s_n + t_n) + (s_n + t_n) - (f + g)|$$

$$\leq \int_I |h - (s_n + t_n)| + \int_I |f - s_n| + \int_I |g - t_n|$$

$$< \frac{\varepsilon}{3} + \frac{\varepsilon}{3} + \frac{\varepsilon}{3} = \varepsilon.$$

Thus, $d(h,(f + g)) = 0$ and $h = (f + g) \in L(I)$. Also, we have

$$\int_I f + g = \int_I h = \int_I \lim_{n \to \infty} (s_n + t_n)$$

$$= \int_I \lim_{n \to \infty} s_n + \int_I \lim_{n \to \infty} t_n = \int_I f + \int_I g.$$

In summary, in this section we have shown that the integral of a Lebesgue function is defined in terms of a limit of the integral of the generating step function sequence (when this limit exists). (In other words, when this integral converged to a finite real number.) Another way to consider this is that a function is Lebesgue integrable if and only if there exists a finite real number M such that

$$\left| \int_I f \right| < M.$$

RIEMANN AND LEBESGUE INTEGRALS

Up to this point we have been discussing the Lebesgue integral from a purely theoretical, or conceptual, point of view. We really haven't said anything about how to actually calculate this integral assuming we were given some particular function (say, for example, $f(x) = x^2$ over the closed interval $I = [0,1]$). Before we address this topic, let's first briefly review the Riemann integral of elementary calculus. The *Riemann integral* is defined, for a function f on a closed interval $[a,b]$, as follows: Let $P = \{x_0, x_1, ..., x_n\}$ be a partition of $[a,b]$ and let $t_k \in [x_{k-1}, x_k]$, $k = 1, 2, ..., n$. We then form the sequence of Riemann sums given as

$$S_n = \sum_{k=1}^{n} f(t_k)(x_k - x_{k-1}). \tag{3.1}$$

Next, we consider the limit of S_n as we take finer and finer partitions and let n approach infinity. If the function is sufficiently well behaved, then there is some hope that S_n will approach a definite limit, which we call the Riemann integral, notationally written as

$$\int_a^b f(x)\,dx.$$

When such a limit does indeed exist, we say the function f is Riemann integrable and write $f \in R([a,b])$. In terms of ε and N, we write $f \in R([a,b])$ if, and only if, given any $\varepsilon > 0$, there exists a positive integer N such that $n > N$ implies

$$\left| S_n - \int_a^b f(x)\,dx \right| < \varepsilon.$$

If we examine equation (3.1) we see that the Riemann integral may also be considered the sum of the area of a sequence of step functions, or rectangles. In this case, $f(t_k)$ is analogous to c_k and describes the value of the step function over the partition subinterval (x_{k-1}, x_k).

When dealing with the Lebesgue integral we first considered convergence of the step function sequence to a limit function f. The existence and convergence of the integral of f followed directly. In other words, when dealing with the Lebesgue theory, if we can show that the sequence of step functions is either convergent or Cauchy convergent, then the integral (of the limit function) will exist.

Using Riemann theory, we begin with a function f and generate the individual rectangles using values of the function within each partition subinterval. In this situation we do not deal with convergence to the function f but, instead, only with the convergence of the Riemann sums of equation (3.1). This requires that we place conditions or restrictions on f to guarantee the existence and convergence of the integral. Normally, continuity and boundedness of f are sufficient conditions to guarantee that the Riemann integral will exist.

Just as with our derivation of the Lebesgue integral, the Riemann integral is of little calculational use to us in this form. The reason that the Riemann integral of a function is so useful is because of the relationship that it has to the derivative of the function. This relationship is clearly stated by the *Fundamental Theorem of Calculus*, which is the link that relates the integral and differential calculus. That is to say, the Riemann integral of a function $f(x)$ is given by its anti-derivative $g(x)$; in other words,

$$\text{if} \quad f(x) = g'(x) = \frac{dg(x)}{dx}, \quad \text{then} \quad \int f(x) = g(x) + C.$$

The next theorem tells us when the Rieman and Lebesgue integrals are equal.

Theorem 3.5. Let f be defined and bounded on a closed interval $[a,b]$. Assume also that f is continuous a.e. on $[a,b]$. Then $f \in L([a,b])$, and the integral of f as a function in $L([a,b])$ is equal to the Riemann integral of f over $[a,b]$. (For the proof of this theorem see problem 8 at the end of this chapter.)

The importance of this theorem cannot be overemphasized. It tells us when the Riemann and Lebesgue integrals are equal and, therefore, tells us when we can use the Fundamental Theorem of Calculus and all of our familiar rules to calculate the Lebesgue integral. For example, the Lebesgue integral of $f(x) = x^2$ on $[a,b]$ is given by

$$\int_I f = \int_a^b x^2 \, dx = \frac{b^3 - a^3}{3}$$

because x^2 is bounded, defined, and continuous over the finite interval $[a,b]$.

CONVERGENCE OF SEQUENCES OF LEBESGUE INTEGRABLE FUNCTIONS

As we soon demonstrate, a real strength of Lebesgue integration theory becomes obvious when dealing with convergence and term-by-term integration of sequences of functions. In this section we present several theorems that deal with convergence of sequences and series of Lebesgue functions.

We first show that $L(I)$ is a complete space, that is to say, all Cauchy sequences of functions $\{f_n\} \in L(I)$ converge to a limit function f which is also in $L(I)$.

Theorem 3.6. The space of Lebesgue integrable functions $L(I)$ is a complete set.

Proof. We must show that all Cauchy sequences of functions $\{f_n\} \in L(I)$ converge to a limit function f, which is an element of $L(I)$. To accomplish this, let us assume that $\{f_n\}$ is a sequence of functions in $L(I)$ which is Cauchy convergent, a.e. on I, thus $\{f_n\}$ converges to some limit function f; that is, given any $\varepsilon > 0$, there exists an integer N such that $n > N$ implies $d(f_n, f) < \varepsilon/2$, a.e. on I.

Now, for each $\{f_n\} \in L(I)$ there is (by definition) a sequence of step functions $\{s_{n,k}\}$ which generates f_n. That is, given any $\varepsilon > 0$ there exists an integer M such that $k > M$ implies $d(s_{n,k}, f_n) < \varepsilon/2$, a.e. on I.

Now, by properties of the norm of $L(I)$, we have, for $n > N$ and $k > M$,

$$d(s_{n,k}, f) \leq d(s_{n,k}, f_n) + d(f_n, f) < \frac{\varepsilon}{2} + \frac{\varepsilon}{2} = \varepsilon.$$

If we choose a sequence of step functions $\{t_m\}$ given by

$$t_m = s_{m,m}, \quad \text{where} \quad m = \text{maximum } (n, k),$$

then we have a sequence $\{t_m\}$ of step functions which generates f and, therefore, by definition $f \in L(I)$. Q.E.D.

Theorem 3.7 (Beppo–Levi Theorem for Sequences). Assume that $\{f_n\}$ is an increasing sequence (a.e. on I) of functions in $L(I)$ and that $\{\int f_n\}$ is bounded from above. Then $\{f_n\}$ converges a.e. on I to a limit function $f \in L(I)$ and

$$\int_I f = \lim_{n \to \infty} \int_I f_n.$$

Proof. Since $\{f_n\}$ is monotonically increasing, a.e. on I, Theorem 3.4-c implies that $\{\int f_n\}$ is also monotonically increasing. Now, $\{\int f_n\}$ increasing and (by assumption) bounded above implies that it is convergent and, thus, a Cauchy sequence; that is, given any $\varepsilon > 0$, there exists an integer N such that $n > N$ and $m > N$ imply

$$\left| \int_I f_n - \int_I f_m \right| < \varepsilon.$$

Now, $\{f_n\}$ being an increasing sequence allows us to write

$$|f_n - f_m| = \pm(f_m - f_n);$$

thus,

$$\int_I |f_m - f_n| = \pm \int_I (f_m - f_n) = \left| \int_I (f_m - f_n) \right| < \varepsilon.$$

Therefore, $\{f_n\}$ is a Cauchy sequence and since $L(I)$ is a complete space, we have $\{f_n\}$ converges to a limit $f \in L(I)$. This fact, in turn, implies

$$\int_I f = \lim_{n \to \infty} \int_I f_n. \qquad \text{Q.E.D.}$$

If we consider the sequence

$$f_n = \sum_{k=1}^{n} g_k,$$

then we can apply Theorem 3.7 to obtain the following theorem.

Theorem 3.8 (Beppo–Levi Theorem for Series). Let $\{g_n\}$ be a sequence of functions in $L(I)$ such that:

   (a) g_n is nonnegative, a.e. on I, and

   (b) $\displaystyle\sum_{n=1}^{\infty} \int_I g_n$ is convergent.

Then

$$\sum_{n=1}^{\infty} g_n \text{ converges, a.e. on } I, \text{ to a sum function } g \in L(I)$$

and

$$\int_I g = \int_I \sum_{n=1}^{\infty} g_n = \sum_{n=1}^{\infty} \int_I g_n.$$

Another version of the Beppo–Levi Theorem for series is given by the following theorem.

Theorem 3.9. Let $\{g_n\}$ be a sequence of functions in $L(I)$ such that

$$\sum_{n=1}^{\infty} \int_I |g_n| \quad \text{is convergent;}$$

then

$$\sum_{n=1}^{\infty} g_n \quad \text{converges, a.e. on } I, \text{ to a sum function } g \in L(I)$$

and

$$\int_I g = \int_I \sum_{n=1}^{\infty} g_n = \sum_{n=1}^{\infty} \int_I g_n.$$

The three theorems of Levi, which we have just presented, give sufficient conditions for convergence of sequences and series of functions. Theorem 3.7 tells us when we can take the limit under the integral sign. Theorems 3.8 and 3.9 tell us when it is permitted to interchange the summation and integral signs. An example problem is in order at this point.

Example 1. Let $f(x) = x^s$ for $x > 0$ and $f(0) = 0$. Show that $f \in L(I)$ and that

$$\int_0^1 f(x)\,dx = \frac{1}{s+1} \qquad \text{for all} \quad s > -1.$$

SOLUTION. If $s \geq 0$, then f is defined, bounded and continuous on $[0,1]$ and thus $f \in R([0,1])$, which implies $f \in L([0,1])$. Its Riemann (and Lebesgue) integral is equal to $1/(s + 1)$. Now, however, if $s < 0$, then f is unbounded on $[0,1]$ (since $f(0) = 1/0$) and, therefore, f is not Riemann integrable over the interval $[0,1]$. Our approach is to find an increasing sequence of functions that are Lebesgue integrable and convergent to f so that we can then apply Theorem 3.7. We choose $\{f_n\}$ as follows:

$$f_n(x) = \begin{cases} x^s, & \text{if } x \geq 1/n, \\ 0, & \text{if } x < 1/n. \end{cases}$$

Obviously, $\{f_n\}$ is increasing and each f_n is Riemann integrable and consequently Lebesgue integrable, on $[0,1]$; that is,

$$\int_0^1 f_n(x)\,dx = \int_{1/n}^1 x^s\,dx = \frac{1}{s+1}\left(1 + \frac{1}{n^{s+1}}\right).$$

Thus, $\{\int f_n\}$ is a sequence of real numbers which converges to $1/(s+1)$ for all $s > -1$, in which case we have

$$\lim_{n \to \infty} f_n = f$$

and

$$\int_I f = \lim_{n \to \infty} \int_I f_n = \frac{1}{s+1}.$$

We now present Lebesgue's dominated convergence theorem for sequences which should be compared to (Levi's) Theorem 3.7. Basically, Theorem 3.7 used the fact that $\{\int f_n\}$ converged to determine convergence of $\{f_n\}$. Lebesgue's theorem, on the other hand, uses the fact that $\{f_n\}$ converges to prove convergence of $\{\int f_n\}$. Both theorems conclude that term-by-term integration of the sequence is valid; that is,

$$\int_I f = \lim_{n \to \infty} \int_I f_n.$$

Theorem 3.10 (Lebesgue Convergence Theorem for Sequences). Let $\{f_n\}$ be a sequence of Lebesgue integrable functions on an interval I, and assume the following:

$$\lim_{n \to \infty} f_n = f, \qquad \text{a.e. on } I.$$

Then $f \in L(I)$, $\{\int f_n\}$ converges, and

$$\int_I f = \lim_{n \to \infty} \int_I f_n.$$

Proof. By assumption the sequence $\{f_n\}$ converges to a limit function f. Since each function f_n is an element of $L(I)$ and $L(I)$ is a complete space, we know that $f \in L(I)$. Also, since $\{f_n\}$ converges, it is a Cauchy sequence. Thus, given any $\varepsilon > 0$, we can always find an N such that for $n > N$ and $m > N$ we have

$$\int_I |f_m - f_n| < \varepsilon.$$

Now let us consider

$$\left| \int_I f_m - \int_I f_n \right| = \left| \int_I (f_m - f_n) \right|$$

$$\leq \int_I |f_m - f_n| < \varepsilon.$$

Thus, $\{\int f_n\}$ forms a Cauchy sequence of real numbers; and since the real number system is a complete space, this sequence converges to the real number $\int f$. These facts imply

$$\int_I f = \lim_{n \to \infty} \int_I f_n. \qquad \text{Q.E.D.}$$

The corresponding Lebesgue theorem for series follows.

Theorem 3.11 (Lebesgue Theorem for Series). Let $\{g_n\}$ be a sequence of functions in $L(I)$ such that:
(a)   each g_n is nonnegative a.e. on I, and
(b)   the series

$$\sum_{n=1}^{\infty} g_n$$

converges a.e. on I to a sum function g which is bounded above by a function in $L(I)$. Then $g \in L(I)$, the series

$$\sum_{n=1}^{\infty} \int_I g_n$$

converges, and

$$\int_I g = \int_I \sum_{n=1}^{\infty} g_n = \sum_{n=1}^{\infty} \int_I g_n.$$

LEBESGUE FUNCTIONS ON AN UNBOUNDED INTERVAL

Thus far, we have only considered integrals over bounded intervals. However, it is often necessary to evaluate an integral over an unbounded interval such as $I = [a, \infty)$. In this section we consider such an integral. Let f be defined on the half-infinite interval $I = [a, \infty)$, and assume f is Lebesgue integrable on a compact interval $[a, b]$, for each $b \geq a$, such that there exists a positive constant M for which

$$\int_a^b |f(x)| \, dx < M \qquad \text{for all} \quad b \geq a.$$

Then $f \in L(I)$, the limit

$$\lim_{b \to \infty} \int_a^b f$$

exists, and

$$\int_a^\infty f \equiv \lim_{b \to \infty} \int_a^b f.$$

We have a corresponding result for the interval $(-\infty, a\}$; that is,

$$\int_{-\infty}^a f = \lim_{b \to -\infty} \int_b^a f,$$

provided that

$$\int_a^b |f(x)| dx < M \qquad \text{for all} \quad b \geq a.$$

Combining the previous two results, we have

$$\int_{-\infty}^\infty f = \lim_{a \to -\infty} \int_a^c f + \lim_{b \to \infty} \int_c^b f.$$

Example 2. Consider the function

$$f(x) = \frac{1}{1 + x^2} \qquad \text{for all} \quad x \in R^1.$$

We first note that $f(x)$ is nonnegative; thus,

$$\int_a^b |f(x)| dx = \int_a^b f(x) dx = \int_a^b \frac{dx}{1 + x^2}$$

$$= \arctan(b) - \arctan(a) \leq \pi.$$

Thus, we have $f \in L(R^1)$ and

$$\int_{-\infty}^\infty f = \lim_{a \to -\infty} \int_a^0 \frac{dx}{1 + x^2} + \lim_{b \to \infty} \int_0^b \frac{dx}{1 + x^2}$$

$$= \lim_{a \to -\infty} \arctan(a) + \lim_{b \to \infty} \arctan(b) = \frac{\pi}{2} + \frac{\pi}{2} = \pi.$$

We recall, from elementary calculus, that the previous integrals look very similar to the improper Riemann integral, which is defined as follows:
If $f \in R([a,b])$ for every $b \geq a$ and if the limit

$$\lim_{b \to \infty} \int_a^b f(x) dx$$

exists, then f is said to be Riemann integrable, on $[a,\infty)$, and the improper integral is defined as

$$\int_a^\infty f(x)\,dx = \lim_{b\to\infty} \int_a^b f(x)\,dx$$

The next theorem tells us when the improper Riemann integral and the Lebesgue integral on an unbounded interval are equal.

Theorem 3.12. Assume $f \in R([a,b])$ for every $b \geq a$ and assume there is a positive constant M such that

$$\int_a^b |f(x)|\,dx \leq M \qquad \text{for all}\quad b \geq a;$$

then both f and $|f|$ are improper Riemann integrals on $[a,\infty)$. Also, $f \in L([a,\infty))$ and the Lebesgue integral of f is equal to the improper Riemann integral of f.

Example 3. Consider the integral

$$\int_1^\infty e^{-x} x^{y-1}\,dx.$$

We first note that

$$\lim_{x\to\infty} e^{-x/2} x^{y-1} = 0;$$

thus, there exists a constant M such that

$$e^{-x/2} x^{y-1} \leq M \qquad \text{for all}\quad x \geq 1.$$

Therefore,

$$e^{-x} x^{y-1} \leq M e^{-x/2},$$

from which we can easily show (for $b \geq 1$)

$$\int_1^b |f(x)|\,dx \leq M \int_1^b e^{-x/2}\,dx,$$

$$\leq M \int_0^b e^{-x/2}\,dx = 2M(1 - e^{-b/2}) \leq 2M.$$

Therefore,

$$\int_1^\infty e^{-x} x^{y-1}\,dx$$

exists for every real y, both as an improper Riemann integral and as a Lebesgue integral on an unbounded interval.

INTEGRAL FUNCTIONS

An integral function $F(y)$ is a function of the form

$$F(y) = \int_X f(x,y)\,dx,$$

where f is a real-valued function of two variables defined on the product set $X \times Y$, in which both X and Y are general subintervals of R^1. Obviously the Fourier transform of a function is an integral function. This can be appreciated by letting

$$f(x,y) = f(x)e^{-2\pi ixy}$$

in the above equation.

In this section we present several useful theorems concerning properties of integral functions. More specifically, we discuss the conditions under which continuity, differentiability, and integrability are transmitted from the integrand function f to the function F. Inasmuch as we do not wish to become excessively embroiled in the details of this topic, we present these theorems without proof and concern ourselves mainly with the overall results and what they mean.

The first theorem discusses continuity and tells us when we can interchange the limit and integral sign.

Theorem 3.13. Let X and Y be two subintervals of R^1 and let f be defined on the product set $X \times Y$, satisfying the following conditions:

(a) For each fixed $y \in Y$, the function f_y defined on Y by the equation

$$f_y(x) = f(x,y)$$

is Lebesgue integrable on X.

(b) There exists a nonnegative function $g \in L(I)$ such that for each $y \in Y$, $f(x,y)$ is dominated a.e. on X by g; that is,

$$|f(x,y)| \leq g(x), \qquad \text{a.e. on } X.$$

(c) For each fixed $y \in Y$ we have

$$\lim_{t \to y} f(x,t) = f(x,y), \qquad \text{a.e. on } X.$$

Then, the Lebesgue integral

$$\int_X f(x,y)\,dx$$

exists for each $y \in Y$, and the function F defined by

$$F(y) = \int_X f(x,y)\,dx$$

is continuous on Y; that is, if $y \in Y$ we have

$$\lim_{t \to y} \int_X f(x,t)dx = \int_X \lim_{t \to y} f(x,t)dx.$$

The main result of this theorem is that if conditions (a)–(c) are satisfied, then we can interchange the limit and integral sign.

The next theorem provides us with the conditions under which interchanging the differential and integral—or, more properly, differentiation under the integral sign—is permitted.

Theorem 3.14. Let X and Y be two general subintervals of R^1 and let f be a function defined on the product set $X \times Y$, satisfying the following conditions:

(a) For each fixed $y \in Y$, the function f_y defined on X by the equation $f_y(x) = f(x,y)$ is Lebesgue integrable on X.
(b) The partial derivative $D_2 f(x,y)$ exists for each interior point of $X \times Y$.
(c) There is a nonnegative function $G \in L(X)$ such that

$$|D_2 f(x,y)| \le G(x) \qquad \text{for all interior points of } X \times Y,$$

Then, the Lebesgue integral

$$\int_X f(x,y)dx$$

exists for every $y \in Y$, and the function defined by

$$F(y) = \int_X f(x,y)dx$$

is differential at each interior point of Y. Moreover, its derivative is given by the formula

$$\frac{dF(y)}{dy} = F'(y) = \int_X D_2 f(x,y)dx.$$

The final theorem spells out the conditions under which we can interchange the order in integration.

Theorem 3.15. Let X and Y be two subintervals of R^1, and let k be a function that is defined, continuous, and bounded on $X \times Y$ (i.e., $|k(x,y)| \le M$) for all $(x,y) \in X \times Y$. Assume $f \in L(X)$ and $g \in L(Y)$; then we have:

(a) For each $y \in Y$, the Lebesgue integral

$$\int_X f(x)k(x,y)dx$$

exists; and the function F, defined on Y by the equation

$$F(y) = \int_X f(x)k(x,y)dx,$$

is continuous on Y.

(b) For each $x \in X$, the Lebesgue integral

$$\int_Y f(y)k(x,y)dy$$

exists; and the function G, defined on X by the equation

$$G(x) = \int_Y g(y)k(x,y)dy,$$

is continuous on X.

(c) The two Lebesgue integrals

$$\int_Y g(y)F(y)dy \quad \text{and} \quad \int_X f(x)G(x)dx,$$

exist and are equal; that is,

$$\int_X f(x)\left[\int_Y g(y)k(x,y)dy\right]dx = \int_Y g(y)\left[\int_X f(x)k(x,y)dx\right]dy.$$

We finish this section by discussing a particular integral function that illustrates the application of the theorems presented. Consider the following integral function:

$$F(y) = \int_0^\infty e^{-xy}\frac{\sin(x)}{x}dx, \quad \text{for} \quad y > 0.$$

We first apply Theorem 3.13 to show that the function $F(y)$ is continuous on the interval $Y = (0,\infty)$. To accomplish this we must show that the function $f(x,y) = e^{-xy}\sin(x)/x$ satisfies parts (a)–(c) of the theorem.

Part (a). For each $y \in Y$, f certainly safisfies the conditions of Theorem 3.12 and, therefore, is both Riemann and Lebesgue integrable.

Part (b). For any $a > 0$, consider the subinterval $Y_a = [a,\infty)$. Over this interval we have

$$|f(x,y)| = |e^{-xy}\sin(x)/x| \le |e^{-xy}| \le e^{-ax},$$

for $x \in [0,\infty)$. Thus, f is dominated by the Lebesgue integrable function $g(x) = \exp[-ax]$ on Y_a for every $a > 0$.

Part (c). For every $y \in Y$, we see that f is continuous as a function of y.

Therefore, we have satisfied the conditions of Theorem 3.13; this implies that $F(y)$ is continuous on Y_a for every $a > 0$ and, thus, $F(y)$ is continuous on $Y = (0,\infty)$.

Next let us consider the limit of $F(y)$ as y approaches infinity. We wish to show

$$\lim_{y \to \infty} F(y) = 0.$$

To accomplish this we let y_n be an increasing sequence of real numbers, such that $y_n \geq 1$ for all n and $y_n \to \infty$ as $n \to \infty$. Thus, if we can show

$$\lim_{n \to \infty} F(y_n) = 0,$$

then we establish our desired result. We begin by defining a sequence of functions on $X = [0, \infty)$ as

$$f_n(x) = e^{-xy_n} \frac{\sin(x)}{x}.$$

Our approach is to apply Theorem 3.10 (Lebesgue Dominated Convergence) to this sequence. To do this we must show that each $f_n(x)$ is Lebesgue integrable on X. This is straightforward since each f_n is Riemann integrable on $[0, b]$ for all $b \geq 0$ and

$$\int_0^b |f_n| < \int_0^b e^{-x} \, dx < 1.$$

Therefore, $f_n \in L(X)$ for all n. Now,

$$\lim_{n \to \infty} f_n = 0 \qquad \text{a.e. on } X$$

and, therefore, f_n converges a.e. on X (in fact, everywhere except 0) to a limit function $f = 0$. Consequently,

$$\int f = \int 0 = 0 = \lim_{n \to \infty} \int f_n,$$

but

$$\int f_n = F(y_n), \qquad \text{so} \quad F(y_n) \to 0 \quad \text{as} \quad n \to \infty.$$

Thus far, we have discussed properties of $F(y)$ but have really not said anything about evaluating the integral itself. We accomplish this by using Theorem 3.14 (differentiation under the integral). We have already shown that $f(x, y)$ is Lebesgue integrable on X and thus part (a) of the theorem is satisfied. To show that parts (b) and (c) are satisfied, we calculate the partial derivative $D_2 f(x, y)$ on $X = [0, \infty)$ and $Y_a = [a, \infty)$; that is,

$$D_2 f(x, y) = -e^{-xy} \sin(x).$$

This function clearly exists and is dominated by the Lebesgue integrable function $f(x) = \exp[-ax]$, for all interior points of $X \times Y_a$. Therefore, by this theorem we can take the derivative under the integral sign to obtain

$$F'(y) = -\int_0^\infty e^{-xy} \sin(x) \, dx.$$

We now note that the integral

$$\int_0^b e^{-xy} \sin(x) \, dx$$

exists as a Riemann integral for all $x \in X$ and $b > 0$. Furthermore, integration by parts yields

$$\int_0^b e^{-xy} \sin(x) \, dx = \frac{e^{-by}[-y\sin(b) - \cos(b)] + 1}{1 + y^2}.$$

Again we are able to use Theorem 3.12 to show that we can take the limit as b approaches infinity in the above equation to obtain

$$\int_0^\infty e^{-xy} \sin(x) \, dx = \frac{1}{1 + y^2}.$$

Thus we have

$$F'(y) = -\frac{1}{1 + y^2}, \qquad y \in Y_a.$$

Integration of the above equation yields

$$F(y) - F(b) = \int_b^y \frac{dt}{1 + t^2} = \arctan(b) - \arctan(y)$$

or

$$F(y) = F(b) + \arctan(b) - \arctan(y).$$

However, in the limit as $b \to \infty$ we have $F(b) = 0$ and $\arctan(b) = \pi/2$. Thus

$$F(y) = \pi/2 - \arctan(y), \qquad y \in Y_a.$$

Summarizing, we have

$$\int_0^\infty e^{-xy} \frac{\sin(x)}{x} \, dx = \frac{\pi}{2} - \arctan(y), \qquad y > 0.$$

SQUARE-SUMMABLE FUNCTIONS

In this section we introduce and briefly discuss square-summable functions. These functions allow us to consider convergence properties of the Fourier series representation of a function.

If $f^2 \in L(I)$, then the nonnegative number, denoted as $\|f\|$ or $m_2(f)$, is called the L^2-norm of f.

We denote by $L^2(I)$ the set of all real valued functions on I such that $f^2 \in L(I)$. The functions in $L^2(I)$ are said to be *square-summable* or *square-integrable*. The following theorems deal with properties of these types of functions.

Theorem 3.21 (Schwartz). Assume $f, g \in L^2(I)$, then

$$\left[\int_I f(t)g(t)\,dt \right]^2 < \int_I f^2(t)\,dt \int_I g^2(t)\,dt.$$

Proof. Consider the following equation:

$$\int_I (f(t) + ag(t))^2\,dt = \int_I f^2(t)\,dt + 2a \int_I f(t)g(t)\,dt + a^2 \int_I g^2(t)\,dt \geq 0,$$

where a is any real constant. For notational convenience we make the following substitutions:

$$A = \int_I f^2(t)\,dt,$$

$$B = \int_I f(t)g(t)\,dt,$$

$$C = \int_I g^2(t)\,dt.$$

The previous equation then becomes

$$A + 2Ba + Ca^2 \geq 0.$$

This can be considered a quadratic equation in a. Since it is always greater than or equal to zero, we know that it cannot have real distinct roots; that is,

$$B^2 - AC \leq 0 \quad \text{or} \quad B^2 \leq AC.$$

When we substitute back for A, B, and C we obtain our desired results.

Q.E.D.

Theorem 3.22. If $f \in L^2(I)$ and $g \in L^2(I)$, then for all real numbers a and b we have:

(a) $f \cdot g \in L^2(I)$ and
(b) $(af + bg) \in L^2(I)$.

Proof. To prove part (a) we must first demonstrate a basic inequality. We proceed as follows:

$$(f(t) - g(t))^2 = f^2(t) + g^2(t) - 2f(t)g(t) \geq 0.$$

Thus

$$f^2(t) + g^2(t) \geq 2f(t)g(t)$$

or

$$f(t)g(t) \leq \tfrac{1}{2}(f^2(t) + g^2(t)).$$

Integration of both sides of the above equation yields

$$\int_I f(t)g(t)\,dt \le \tfrac{1}{2}\int_I f^2(t)\,dt + \tfrac{1}{2}\int_I g^2(t)\,dt.$$

By assumption, both integrals on the right-hand side exist (and are bounded). Consequently, the integral on the left-hand side is bounded and exists; that is, $fg \in L^2(I)$.

The proof of part (b) follows directly from part (a); that is,

$$\int_I (f(t)+g(t))^2\,dt = \int_I f^2(t)\,dt + 2\int_I f(t)g(t)\,dt + \int_I g^2(t)\,dt,$$

and each of the integrals on the right-hand side of the above equation exists.

Q.E.D.

Theorem 3.23. The space $L^2(I)$ is complete in the sense of the L^2-norm of I.

SUMMARY

In this chapter we examined the basic theory of integration from the Lebesgue point of view. This was accomplished by introducing the idea of a step function and its integral. After discussing various properties of step functions, we considered sequences of step functions and their integrals. We saw that the set made up of all step functions formed a vector space over the real field. For this vector space we were able to define a norm (in terms of the integral); consequently, we were able to talk about convegence in this space. We showed that if a sequence of step functions converged, then the sequence of real numbers formed by the integral of the step-function sequence terms also converged.

When a sequence of step functions converged, we called the limit function a *Lebesque integrable function*. As it turned out, this limit function may or may not be another step function. Finally, we defined the integral of this limit function as the limit of the individual sequence integrals; that is,

$$\int_I f = \lim_{n\to\infty}\int_I s_n.$$

After discussing properties (such as linearity and order) of Lebesgue functions we considered convergence of sequences and series of Lebesgue functions. As we saw, considering integration in this fashion was particularly convenient for discussing convergence and term-by-term integration of sequences and series of Lebesgue functions.

When it actually came down to evaluating the Lebesgue integral expressions we reverted back to showing when the Lebesgue and Riemann integrals were equal (Theorems 3.5 and 3.12) and then used the Fundamental Theorem of Calculus. In other words, the Lebesgue approach to integration provides theoretical insight into the behavior and existence questions of integrals but does not provide any calculational advantages. As a matter of fact, in this text, we always deal within the realm where the Riemann and Lebesgue integrals are equal.

Finally, in this chapter, we expanded upon our Lebesgue theory to introduce integral and square-summable functions which will be useful to us later in this text.

PROBLEMS

1 Find a Lebesgue covering of
  (a) $(1,2)$
  (b) $[1,2)$
  (c) $(1,3) \cup (5,7) \cup (8,12)$
  (d) $(1,3) \cup (3,4)$.

2 Give an example of a set of measure zero in R^1, R^2, and R^3.

3 Give an example of a property that holds almost everywhere (a.e.) in R^1 and R^2.

4 Prove Theorem 3.1, parts (b), (c), and (d).

5 Show that the set of all step functions forms a vector space over the real field R^1.

6 Show that

$$m(V_s) = \int_I |s| = \sum_{k=1}^{n} |c_k|(x_k - x_{k-1})$$

forms a norm for the vector space of step functions over the real field.

7 Prove Theorem 3.4, parts (b)–(d).

8 From intermediate calculus we know that if a function $f(t)$ is defined, bounded, and continuous on a closed interval $[a,b]$, then its Riemann integral exists. Use this fact, and choose t_k [see equation 3.1] in such a way that $f(t_k) = c_k$; then form a sequence of step functions that will converge to $f(t)$, which guarantees the existence of the Lebesgue integral of f. In this way, Theorem 3.5 is proved.

9 Prove Theorem 3.8 (let $f_n = \sum g_n$ and use Theorem 3.7).

10 Prove Theorem 3.9.

11 Show that if the sequence of functions $f_n \in L(I)$ converges to the function $f \in L(I)$ and the sequence of real numbers $\int f_n$ converges to $\int f$, then

$$\int_I f = \lim_{n \to \infty} \int_I f_n.$$

BIBLIOGRAPHY

Apostal, T. M., *Mathematical Analysis* (second edition), Addison- Wesley, Reading, Mass., 1974.

Arsac, J., *Fourier Transforms and the Theory of Distributions*, Prentice-Hall, Englewood Cliffs, N.J., 1966.

Boas, R. P., *A Primer of Real Functions*, Carus Monograph No. 13, John Wiley & Sons, New York, 1960.

Halmos, P. R., *Finite Dimensional Vector Spaces* (second edition), D. Van Nostrand, Princeton, N.J., 1958.

Hasser, N. B., J. P. Lasalle, and J. A. Sullivan, *Intermediate Analysis*, Blaisdell, New York, 1964.

Rudin, W., *Principles of Mathematical Analysis* (second edition), McGraw-Hill, New York, 1964.

Titchmarsh, E. C., *Fourier Transforms*, Clarendon Press, Oxford, 1948.

4

DISTRIBUTION THEORY

In this chapter we present the basic idea and concept of a distrubution. The main reason for studying distributions is that they allow us to properly discuss such "maverick" functions as the delta and comb. In addition, distribution theory provides us with a means to establish existence and reciprocity of the Fourier transform. In this chapter we use the concept of an inner product as a guide from which we define a distribution.

INNER PRODUCT OF TWO FUNCTIONS

Given two (perhaps complex) functions $f(x)$ and $g(x)$, consider the integral

$$\int_{-\infty}^{\infty} f(x)g(x)dx.$$

When, and if, this integral exists we call it the *inner product* (or scalar product) of f and g, and notationally we write

$$\langle f(x), g(x) \rangle = \int_{-\infty}^{\infty} f(x)g(x)dx. \tag{4.1}$$

This concept of an inner product will be invaluable to us when we define a distribution later in this chapter. Therefore, we now spend a little time looking at some of its properties and consequences.

We first note that the inner product is a number (perhaps complex) that results from performing the integration of equation (4.1).

We also note that the inner product is only defined if the integral of equation (4.1) exists. Certainly if the product function $f(x)g(x)$ is Lebesgue integrable, then we can easily establish existence (based on the results presented in Chapter 3).

Let us now consider several elementary properties of the inner product which follow directly as a result of the properties of the involved integrals. These properties are especially important because they will be used as a guide to define properties of a distribution. (Note that in the following development we have dropped the limits of integration for convenience and ease of presentation. Unless otherwise specifically stated, all integrals are performed from $-\infty$ to $+\infty$).

A. Multiplication by a Constant

$$\langle af(x),g(x)\rangle = a\langle f(x),g(x)\rangle = \langle f(x),ag(x)\rangle$$

B. Commutativity

If the inner product of f and g exists, then we have

$$\langle f(x),g(x)\rangle = \int f(x)g(x)dx = \int g(x)f(x)dx = \langle g(x),f(x)\rangle.$$

C. Distributivity with Respect to Addition

If the inner products of f_1 with g and f_2 with g exist, then we have

$$\langle af_1(x)+bf_2(x),g(x)\rangle = a\langle f_1(x),g(x)\rangle + b\langle f_2(x),g(x)\rangle.$$

D. Associativity

If the inner product of $f(x)g(x)$ with $h(x)$ exists, then we have

$$\langle f(x)g(x),h(x)\rangle = \int f(x)g(x)h(x)dx = \langle f(x),g(x)h(x)\rangle.$$

E. Translation of One Function

If the inner product of f with g exists, then we have

$$\langle f(x-a),g(x)\rangle = \int f(x-a)g(x)dx.$$

If we now let $y = x - a$ (or $x = y + a$), we obtain

$$\langle f(x-a),g(x)\rangle = \int f(y)g(y+a)dy = \langle f(x),g(x+a)\rangle.$$

F. Scale Change of One Function

If the inner product of f with g exists, then we have

$$\langle f(ax), g(x) \rangle = \int f(ax)g(x)dx.$$

If we now let $y = ax$ or $x = y/a$ and $dx = dy/a$, we find

$$\langle f(ax), g(x) \rangle = \int \left(\frac{1}{a}\right) f(y)g\left(\frac{y}{a}\right) dy = \left(\frac{1}{|a|}\right) \left\langle f(x), g\left(\frac{x}{a}\right) \right\rangle.$$

EXISTENCE OF THE INNER PRODUCT

We have already made note of the fact that if the product of $f(x)$ and $g(x)$ is a Lebesgue integrable function i.e., $f \cdot g \in L(I)$, then the inner product as defined by the integral expression of equation (4.1) will indeed exist. We note that this is a sufficient, but not a necessary, condition. In other words, if $f \cdot g \in L(I)$, then the inner product can be guaranteed to exist. However, there may be other functions such that $f \cdot g \notin L(I)$ but still the integral exists. In this section we present additional sufficient condition that guarantee the existence of the inner product.

Let us consider three sets of vector spaces S, D, and C defined in such a way that when $g \in S$ and $f \in D$, their inner product exists and yields an element of the space C. These three sets are described in detail as follows:

C is the vector space of complex numbers over the real field, or simply the complex number system.

S is the vector space of functions g that satisfy the following two conditions:

(1) $g(x)$ is indefinitely differentiable; that is, $g^{[n]}(x)$ exists for any n and all x. (Note that $[n]$ denotes the nth derivative of g.)
(2) $g(x)$ decreases at infinity faster than any power of $1/x$; that is,

$$\lim_{x \to \infty} x^n g(x) = 0 \qquad \text{for all } n.$$

(This limit is taken in the sense of the absolute value norm). An example of this type of function is $g(x) = e^{-|x|}$.

D is the vector space of locally summable functions f, which increase at infinity slower than some power of x. That is to say, there exists some real number X such that

$$\int_{-X}^{X} |f(x)|dx \text{ is bounded,}$$

and there exists some integer n such that

$$\lim_{x \to \infty} \frac{f(x)}{x^n} = 0$$

in the sense of the absolute value norm. $f(x) = x^2$ is an example of a function in this space.

Let's examine the space D a little closer. If $f \in D$, then by definition we can always find a value of n such that

$$\lim_{x \to \infty} \frac{f(x)}{x^n} = 0 ;$$

or, more formally, given any $\varepsilon > 0$ there exists an integer n and a positive real number X such that

$$|f(x)| < \varepsilon |x^n| \qquad \text{for all} \quad x \in ((-\infty, -X) \cup (X, \infty)),$$

or, equivalently, $|x| > X$.

A closer look at the space S reveals that for any function $g(x) \in S$ we have, given any $\varepsilon > 0$ and any positive integer n, we can always find a positive real number X such that

$$|x^n g(x)| < \varepsilon \qquad \text{for} \quad |x| > X.$$

Now if we let $n = m + 2$ (where m is any integer), then by definition

$$|x^m x^2 g(x)| < \varepsilon$$

or

$$|x^m g(x)| < \frac{\varepsilon}{|x^2|} \qquad \text{for} \quad x \in ((-\infty, -X) \cup (X, \infty)).$$

In terms of the sets S, D, and C, the inner product may be considered a law of external composition which takes an element (function) from the set S and combines it with an element from the set D, to yield a third element (complex number) from the set C. The way in which these sets are defined guarantees that the inner product will always exist. This is easily demonstrated as follows:

$$\left| \int_{-\infty}^{\infty} f(x) g(x) \, dx \right| \leq \int_{-\infty}^{\infty} |f(x) g(x)| \, dx$$

$$\leq \int_{-\infty}^{-X} |f(x) g(x)| \, dx + \int_{-X}^{X} |f(x) g(x)| \, dx$$

$$+ \int_{X}^{\infty} |f(x) g(x)| \, dx.$$

Now since all derivatives of $g(x)$ exist (and in particular, $g'(x)$), then $g(x)$ is bounded over the interval $(-X, X)$ and, thus,

$$\int_{-X}^{X} |f(x) g(x)| \, dx \leq \int_{-X}^{X} M' |f(x)| \, dx < M'M.$$

Now given any ε such that $1 > \varepsilon > 0$, we can always choose $X > 1$ such that

$$\int_X^\infty |f(x)g(x)|\,dx \le \int_X^\infty |\varepsilon x^n g(x)|\,dx$$

$$\le \int_X^\infty \frac{\varepsilon^2}{x^2}\,dx = \frac{\varepsilon^2}{X} < \varepsilon.$$

Similar reasoning yields

$$\int_{-\infty}^{-X} |f(x)g(x)|\,dx < \varepsilon.$$

Now combining all three results, we see

$$\int_{-\infty}^\infty |f(x)g(x)\,dx| \le M'M + 2\varepsilon.$$

Thus we have, in effect, shown that for large values of x, the function $g(x)$ decreases faster than $f(x)$ increases. More figuratively, $g(x)$ will eat up $f(x)$ for large values of x, and this guarantees that the inner product integral exists.

DEFINITION OF A DISTRIBUTION

In the previous section we considered the inner product of two functions to be a law of external composition which combined elements (functions) from the sets D and S to obtain an element (complex number) from the set C. Let's now look at this inner product from a slightly different point of view. We consider f to be a mapping of the function $g(x) \in S$ to a complex number. That is to say, the inner product defines the way in which f maps functions from the space S to the set of complex numbers C. For example, let $f(x) = 1$ and $g(x) = \exp[-\pi x^2]$; then

$$\langle f(x), g(x) \rangle = \int_{-\infty}^\infty e^{-\pi x^2}\,dx = 1.$$

Thus, f may be considered to map the function $g(x) = \exp[-\pi x^2]$ to the complex number $(1,0) = 1$. As another illustration, we can easily verify (via an integral table) that f maps the function $g(x) = \exp[-x^2]\cos(x)$ to the number $(\pi e^{-1/2})^{1/2} = 1.38$, that is,

$$\langle f(x), g(x) \rangle = \int_{-\infty}^\infty e^{-x^2}\cos(x)\,dx = 1.38.$$

We now expand this concept and define a distribution as follows: A *distribution* $t(x)$ is a continuous linear mapping from the space S to the set of complex numbers C. We use the notation

$$\langle t(x), g(x) \rangle = z.$$

It is important to note that $t(x)$ does not necessarily denote a function that associates a value $t(x)$ to the real number x. Instead, it designates an operator that, when applied to a function $g(x) \in S$, associates it with a complex number z. Let's look closer at the definition and state clearly what we mean by linear and continuous. When we say $t(x)$ is a linear mapping we mean that for any $g_1(x) \in S$ and $g_2(x) \in S$, we have

$$\langle t(x), ag_1(x) + bg_2(x) \rangle = a\langle t(x), g_1(x) \rangle + b\langle t(x), g_2(x) \rangle,$$

where a and b are any complex numbers. When we say $t(x)$ is a continuous mapping we mean that if

$$\lim_{n \to \infty} g_n(x) = g(x),$$

then

$$\lim_{n \to \infty} \langle t(x), g_n(x) \rangle = \langle t(x), g(x) \rangle.$$

We have purposely chosen our notation so that a distribution applied to a function $g \in S$ will look like an inner product. At this point it is strictly notation, which means the distribution t operates on a function $g \in S$ to yield a complex number z. Thus, when we see $\langle f(x), g(x) \rangle$, it may mean that f is a function such that its inner product with the function $g \in S$ exists and yields a complex number z, or it may mean that $f(x)$ is a distribution which operates on $g \in S$ (in some specific way) to yield a complex number z. When it is necessary to distinguish between the two, we call $\langle t(x), g(x) \rangle$ the distribution inner product. It is interesting to note that all functions f in the previously defined vector space D are linear continuous mappings from S to C and, therefore, satisfy the definition of a distribution. That is to say, *all functions in D are also distributions*. However all distributions are not necessarily in the space D. When, in fact, the distribution $t(x)$ is also a function in the space D we have the mapping rule clearly spelled out (by the definition of the inner product). In other words, when $t(x) \in D$ the mapping rule requires us to first multiply $t(x)$ and $g(x)$ together and then integrate the resulting function from $-\infty$ to ∞.

We now present several very useful distributions. To every function $g(x) \in S$, we *define* the *delta distribution* as that which maps $g(x)$ to $g(0)$ (where $g(0)$ is obviously the value of the function $g(x)$ at $x = 0$); that is,

$$\langle \delta(x), g(x) \rangle = g(0). \tag{4.2}$$

Again we stress the fact that $\delta(x)$ is not a function of x but is, instead, a mapping from S to C. Next we define the distribution $\delta(x, a)$ as that which maps $g(x)$ to $g(a)$, that is,

$$\langle \delta(x, a), g(x) \rangle = g(a). \tag{4.3}$$

The *null distribution* $N(x)$ is, by *definition*, that distribution which maps any function $g(x) \in S$ to the complex zero element, that is,

$$\langle N(x), g(x) \rangle = (0, 0) = 0. \tag{4.4}$$

Finally we present the *derivative distribution* $D_1(x)$ which maps $g(x)$ to $g'(0)$, that is,

$$\langle D_1(x), g(x) \rangle = g'(0). \tag{4.5}$$

Again we restate the important fact that these distributions $\delta(x, a)$ and $D_1(x)$ are not functions that assign a value to x but rather they are to be viewed as mappings from S to C.

PROPERTIES OF DISTRIBUTIONS

In this section we discuss several fundamental properties of distributions. When we discussed the inner product of two functions, earlier in this chapter, we presented several interesting properties such as scale change and translation of one of the functions. We now proceed to arrive at these same results for a distribution operating on a function. Many of these properties are simply defined using the inner product results as a guide.

Distributivity with Respect to Addition

In the definition of a distribution we require that it be a linear mapping from the space S to the space C. This implies that for any two functions $g_1(x)$ and $g_2(x) \in S$ and the distribution $t(x)$ we have

$$\langle t(x), g_1(x) + g_2(x) \rangle = \langle t(x), g_1(x) \rangle + \langle t(x), g_2(x) \rangle.$$

Sum of Two Distributions

We denote the sum of two distributions $t_1(x)$ and $t_2(x)$ as $t_1(x) + t_2(x)$ and define the distribution inner product of this sum with any function $g(x) \in S$ as

$$\langle t_1(x) + t_2(x), g(x) \rangle = \langle t_1(x), g(x) \rangle + \langle t_2(x), g(x) \rangle. \tag{4.6}$$

That is to say, if $t_1(x)$ maps $g(x)$ to the complex number z_1 and $t_2(x)$ maps $g(x)$ to the complex number z_2, then, by definition, $t_1(x) + t_2(x)$ will map $g(x)$ to the complex number $z_1 + z_2$.

Product of a Distribution with a Complex Number

Let's now consider the product of a complex number a with the distribution $t(x)$, that is, $at(x)$. We define the distribution inner product of $at(x)$ with $g(x) \in S$ as

$$\langle at(x), g(x) \rangle = \langle t(x), ag(x) \rangle.$$

Thus, from linearity considerations we find

$$\langle at(x), g(x) \rangle = a \langle t(x), g(x) \rangle. \tag{4.7}$$

Difference of Two Distributions

If we now choose $a = -1$ in equation (4.7), then we easily establish the difference of two distributions $t_1(x)$ and $t_2(x)$ as

$$t_1(x) - t_2(x) = t_1(x) + (-1)t_2(x),$$

and, for any $g(x) \in S$, we have (using equations (4.6) and (4.7))

$$\langle t_1(x) - t_2(x), g(x) \rangle = \langle t_1(x), g(x) \rangle - \langle t_2(x), g(x) \rangle.$$

Equality of Two Distributions

We consider two distributions $t_1(x)$ and $t_2(x)$ to be equal if and only if their difference is the null distribution, that is,

$$\langle t_1(x) - t_2(x), g(x) \rangle = 0$$

or

$$\langle t_1(x), g(x) \rangle = \langle t_2(x), g(x) \rangle.$$

To say it another way, two distributions are equal if and only if they both map $g(x) \in S$ to the same complex number. As an example, let us now look at the special situation in which the distributions are also functions in the vector space D. In this case we have

$$\langle f_1(x) - f_2(x), g(x) \rangle = \int_{-\infty}^{\infty} (f_1(x) - f_2(x))g(x)\,dx = 0.$$

The above integral is taken in the Lebesgue sense; thus, equality means that the functions (distributions) $f_1(x)$ and $f_2(x)$ are equal almost everywhere over the real line.

Product of a Distribution and a Function

When we dealt with functions, we were able to show that the inner product (when it existed) was associative, that is,

$$\langle f(x)g(x), h(x) \rangle = \langle f(x), g(x)h(x) \rangle.$$

We now establish similar results for the distribution inner product and, in doing so, also define what we mean by the product of a function and a distribution. Since distributions are defined as mappings that operate on functions $g \in S$, when we talk about the product of a function $h(x)$ and a distribution $t(x)$ we must define how this product $t(x)h(x)$ will operate on functions $g(x) \in S$. By definition, we say

$$\langle t(x)h(x), g(x) \rangle = \langle t(x), h(x)g(x) \rangle. \tag{4.8}$$

Thus, we see that the product of a function $h(x)$ and a distribution $t(x)$ operating on a function $g(x) \in S$ is, by definition, the same as the distribution

$t(x)$ operating on the product of $h(x)$ and $g(x)$, whenever this product makes sense. An interesting consequence of this definition is now presented. If, for any function $h(x)$, we have $h(0) = 0$ (e.g., $h(x) = x$ or $h(x) = \sin x$), then the product $\delta(x)h(x)$ turns out to be the null distribution. This is easily verified by noting

$$\langle \delta(x)h(x), g(x) \rangle = \langle \delta(x), h(x)g(x) \rangle = h(0)g(0) = 0.$$

Translation of a Distribution

For functions, we have previously demonstrated that

$$\langle f(x-a), g(x) \rangle = \langle f(x), g(x+a) \rangle.$$

Using this as a guide we define the translation of a distribution $t(x-a)$ as

$$\langle t(x-a), g(x) \rangle = \langle t(x), g(x+a) \rangle. \tag{4.9}$$

Thus we see that, by definition, the distribution $t(x-a)$ maps the function $g(x) \in S$ to the same complex number that the distribution $t(x)$ maps $g(x+a) \in S$. Let's apply this to the delta distribution

$$\langle \delta(x-a), g(x) \rangle = \langle \delta(x), g(x+a) \rangle = g(a).$$

We should recognize this as the previously defined distribution $\delta(x, a)$ (equation (4.3)), which is, therefore, only a translation of the Dirac distribution.

Scale Change of a Distribution

For functions, we have

$$\langle f(ax), g(x) \rangle = \frac{1}{|a|} \left\langle f(x), g\left(\frac{x}{a}\right) \right\rangle,$$

where a is a real number. Thus for distributions, we are motivated to define

$$\langle t(ax), g(x) \rangle = \frac{1}{|a|} \left\langle t(x), g\left(\frac{x}{a}\right) \right\rangle.$$

For the special case when $a = -1$ we have

$$\langle t(-x), g(x) \rangle = \langle t(x), g(-x) \rangle.$$

THE COMB DISTRIBUTION

In this section we introduce the comb distribution, which will be extremely useful to us when we discuss sampling theory in Chapter 9. A comb distribution of interval width Δx (denoted as $\text{comb}_{\Delta x}(x)$) is, by definition, the distribution that maps every function $g(x) \in S$ to the infinite sequence $\{g_k\}$ whose

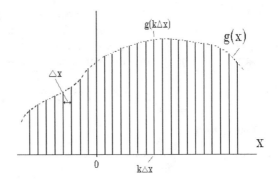

Figure 4.1 Graphical display of a sequence.

terms are given as $g_k = g(k\Delta x)$, that is,

$$\langle \text{comb}_{\Delta x}(x), g(x) \rangle = \{g(k\Delta x)\}, \qquad k \in (-\infty, \infty). \tag{4.10}$$

In other words, a comb distribution converts a function $g(x) \in S$ to a sequence. In Chapter 2 we saw that an infinite sequence is simply a function whose domain is the set of all integers. In equation (4.10) each term $g(k\Delta x)$ may indeed be considered a term of a sequence whose domain is the set of all integers and whose range is given by the set of terms whose values are given by values of the function at $x = k\Delta x$. Just as with any function we can draw a graph of $g(k\Delta x)$ vs. x (or $k\Delta x$) such as the example shown in Figure 4.1.

Next let us consider an individual term of the sequence $g_k = g(k\Delta x)$ in terms of the delta distribution, that is,

$$g(k\Delta x) = \langle \delta(x - k\Delta x), g(x) \rangle.$$

Thus, we see that each individual term in the sequence can be described in terms of a shifted delta distribution. From this we conclude

$$\langle \text{comb}_{\Delta x}(x), g(x) \rangle = \cdots + \langle \delta(x + 2\Delta x), g(x) \rangle + \langle \delta(x + \Delta x), g(x) \rangle$$
$$+ \langle \delta(x), g(x) \rangle + \langle \delta(x - \Delta x), g(x) \rangle$$
$$+ \langle \delta(x - 2\Delta x), g(x) \rangle + \cdots$$

$$\langle \text{comb}_{\Delta x}(x), g(x) \rangle = \sum_{k=-\infty}^{\infty} \langle \delta(x - k\Delta x), g(x) \rangle = \left\langle \sum_{k=-\infty}^{\infty} \delta(x - k\Delta x), g(x) \right\rangle.$$

Thus,

$$\text{comb}_{\Delta x}(x) = \sum_{k=-\infty}^{\infty} \delta(x - k\Delta x). \tag{4.11}$$

We see, therefore, that the comb distribution is in fact a train of equally spaced (Δx apart) delta distributions. This is illustrated graphically in Figure 4.2.

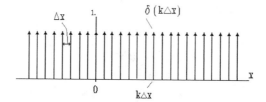

Figure 4.2 The comb distribution.

In summary, the comb distribution of interval width Δx is the distribution that maps any function $g(x) \in S$ to a sequence $\{g_k\}$ whose individual terms are given as $g_k = g(k \Delta x)$. We have also showed that the comb distribution can be considered to be an infinite train of equally spaced unit impulse distributions.

DERIVATIVE OF A DISTRIBUTION

In this section we define what is meant by the derivative of a distribution. Given a distribution $t(x)$, we define its derivative (denoted as $t'(x)$) as that distribution which maps $g(x)$ to the negative value of the complex number to which $t(x)$ maps $g'(x)$. Mathematically, we have

$$\langle t'(x), g(x) \rangle = -\langle t(x), g'(x) \rangle.$$

Higher derivatives are similarly defined as

$$\langle t^{[n]}(x), g(x) \rangle = -\langle t(x), g^{[n]}(x) \rangle.$$

Let's first consider the special situation in which the distribution is also a function f in the vector space D which is continuous and differentiable. Certainly, in this situation we must assume that the mapping from S to C is accomplished via the inner product; that is, we consider

$$\langle f'(x), g(x) \rangle = \int_{-\infty}^{\infty} f'(x) g(x) \, dx.$$

We easily use integration by parts to obtain

$$\int_{-\infty}^{\infty} f'(x) g(x) \, dx = f(x) g(x) \Big|_{-\infty}^{\infty} - \int_{-\infty}^{\infty} f(x) g'(x) \, dx.$$

However, the product $f(x) g(x)$ vanishes at both $+\infty$ and $-\infty$ (why?), and thus we have

$$\int_{-\infty}^{\infty} f'(x) g(x) \, dx = - \int_{-\infty}^{\infty} f(x) g'(x) \, dx$$

or, in distribution notation,

$$\langle f'(x), g(x) \rangle = -\langle f(x), g'(x) \rangle.$$

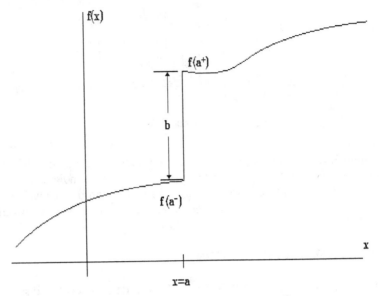

Figure 4.3   Derivative of a function with a jump discontinuity.

Thus we see that if $t(x)$ is also a function and has a derivative in the function sense, then the derivative of $t(x)$ as a distribution is the same as the derivative of $t(x)$ as a function. It is constructive at this point to consider the derivative of a function $f \in D$ with a jump discontinuity. That is, we consider a function $f \in D$ which is everywhere continuous and differentiable except at the point a where it has a jump discontinuity of magnitude b (see Figure 4.3), that is,

$$\lim_{\varepsilon \to 0}[f(a+\varepsilon) - f(a-\varepsilon)] = b.$$

We proceed by considering the inner product of $f(x)$ with a function $g'(x) \in S$:

$$\int_{-\infty}^{\infty} f(x)g'(x)\,dx = \lim_{\varepsilon \to 0}\left(\int_{-\infty}^{a-\varepsilon} f(x)g'(x)\,dx + \int_{a+\varepsilon}^{\infty} f(x)g'(x)\,dx\right).$$

Now, by assumption, $f(x)$ is well behaved over $(-\infty, a) \cup (a, \infty)$; therefore we can use integration by parts to obtain

$$\int_{-\infty}^{a-\varepsilon} f(x)g'(x)\,dx = f(x)g(x)\Big|_{-\infty}^{a-\varepsilon} - \int_{-\infty}^{a-\varepsilon} f'(x)g(x)\,dx$$

$$= f(a-\varepsilon)g(a-\varepsilon) - \int_{-\infty}^{a-\varepsilon} f'(x)g(x)\,dx.$$

(Note: $f(\infty)g(\infty) = 0$.) Similarly, we have

$$\int_{a+\varepsilon}^{\infty} f(x)g'(x)\,dx = -f(a+\varepsilon)g(a+\varepsilon) - \int_{a+\varepsilon}^{\infty} f'(x)g(x)\,dx.$$

Combining the above equations we arrive at

$$\int_{-\infty}^{\infty} f(x)g'(x)\,dx = \lim_{\varepsilon\to 0}\{f(a-\varepsilon)g(a-\varepsilon) - f(a+\varepsilon)g(a+\varepsilon)\}$$

$$- \lim_{\varepsilon\to 0}\left\{\int_{-\infty}^{a-\varepsilon} f'(x)g(x)\,dx + \int_{a+\varepsilon}^{\infty} f'(x)g(x)\,dx\right\}.$$

Now since $g(x)$ is continuous at a we can write

$$\lim_{\varepsilon\to 0} g(a-\varepsilon) = \lim_{\varepsilon\to 0} g(a+\varepsilon) = g(a).$$

Thus,

$$\lim_{\varepsilon\to 0}[f(a-\varepsilon)g(a-\varepsilon) - f(a+\varepsilon)]g(a+\varepsilon)$$

$$= \lim_{\varepsilon\to 0}[f(a-\varepsilon) - f(a+\varepsilon)]g(a) = -bg(a).$$

We also note that since the point $x = a$ forms a set of measure zero and since the above integrals are taken in the Lebesgue sense, we have

$$\lim_{\varepsilon\to 0}\left\{\int_{-\infty}^{a-\varepsilon} f'(x)g(x)\,dx + \int_{a+\varepsilon}^{\infty} f'(x)g(x)\,dx\right\} = \int_{-\infty}^{\infty} f'(x)g(x)\,dx$$

or

$$\int_{-\infty}^{\infty} f(x)g'(x)\,dx = -bg(a) - \int_{-\infty}^{\infty} f'(x)g(x)\,dx.$$

We now write (the number) $bg(a)$ as $\langle b\delta(x-a),g(x)\rangle$, and thus we are able to express the above equation as

$$\langle f(x),g'(x)\rangle = -\langle b\delta(x-a),g(x)\rangle - \langle f'(x),g(x)\rangle,$$
$$\langle f(x),g'(x)\rangle = -\langle b\delta(x-a) + f'(x),g(x)\rangle$$
$$= -\langle t'(x),g(x)\rangle,$$

where

$$t'(x) = b\delta(x-a) + f'(x).$$

Thus we consider $t'(x)$ to be the distribution that is the derivative of a function $f \in D$ which is continuous and differentiable everywhere except at the point $x = a$, where it has a jump discontinuity of magnitude b. It turns out that $t'(x)$ is equal to the derivative of $f(x)$ in the function sense everywhere except at $x = a$. At this point, $t'(x)$ is equal to $b\delta(x-a)$.

ODD AND EVEN DISTRIBUTIONS

Just as we have odd and even functions, we also have odd and even distributions. Consider the following definitions:

A distribution $t(x)$ is called *even* if and only if for every $g(x) \in S$ we have

$$\langle t(x), g(-x) \rangle = \langle t(x), g(x) \rangle.$$

A distribution is called *odd* if and only if for every $g(x) \in S$ we have

$$\langle t(x), g(-x) \rangle = -\langle t(x), g(x) \rangle.$$

Let's consider the following interesting result. Assume that $t(x)$ is an even distribution and that $g(x)$ is an odd function. Then if we first use the properties of the odd function $g(x)$, we find

$$\langle t(x), g(-x) \rangle = \langle t(x), -g(x) \rangle = -\langle t(x), g(x) \rangle.$$

Now using the properties of the even distribution $t(x)$, we obtain

$$\langle t(x), g(-x) \rangle = \langle t(x), g(x) \rangle;$$

thus, combining the above two equations, we see $\langle t(x), g(x) \rangle = 0$. The same result is true if t is an odd distribution and g is an even function.

We end this section with the following three definitions. A distribution $t(x)$ is called *real* if and only if for every real function $g(x) \in S$ we have $\langle t(x), g(x) \rangle$ is a real number. A distribution is called *pure imaginary* if and only if for every real function $g(x) \in S$ we have $\langle t(x), g(x) \rangle$ is an imaginary number. Given a distribution $t(x)$, the *complex conjugate* of $t(x)$, denoted as $t^*(x)$, is the distribution such that

$$\langle t^*(x), g(x) \rangle = [\langle t(x), g^*(x) \rangle]^*.$$

THE DELTA DISTRIBUTION: AN INTUITIVE APPROACH

Thus far in this chapter we have considered the formal definition and properties of distributions. One of the main reasons for introducing distributions is to properly describe the "maverick" or ill-behaved delta and comb functions. Inasmuch as these functions[†] are so useful and important in the study of physical phenomena, we end this chapter with a more intuitive (and certainly less rigorous) examination of the delta function. The purpose of this final section is to translate some of the rigorous distribution results from the purely theoretical to a more "practical" point of view.

In Chapter 1 we considered the delta function to be defined as the limiting case of a sequence of pulse functions of unit area. In this way we came to

---

[†]In this section, for convenience and to keep with the spirit of an intuitive approach we refer to the delta and comb distributions as the delta and comb functions.

interpret a delta function as an infinitesimally brief pulse of infinite amplitude. By definition, the delta distribution is given as follows:

$$\langle \delta(x), g(x) \rangle = g(0).$$

If $\delta(x)$ were a function (which it really isn't), then we would use the inner product integral of equation (4.1) to describe it, that is,

$$\int_{-\infty}^{\infty} \delta(x) g(x) dx = g(0).$$

This above integral can be interpreted as $\delta(x) = 0$ a.e. on $(-\infty, \infty)$. More specifically, it is equal to zero everywhere except at $x = 0$. Mathematically,

$$\int_{-\infty}^{\infty} \delta(x) g(x) dx = \int_{0-}^{0+} \delta(x) g(x) dx = g(0).$$

Since $g(x)$ is continuous and assumed well behaved over the small interval $(0^-, 0^+)$, we can assume that it is a constant equal to the value $g(0)$. Consequently, it can be removed from under the integral sign, that is,

$$g(0) \int_{0-}^{0+} \delta(x) dx = g(0) \quad \text{or} \quad \int_{0-}^{0+} \delta(x) dx = 1.$$

In other words, the area of a delta function taken over the infinitesimal interval $(0^-, 0^+)$ is equal to unity. This is consistent with the way in which the delta function was presented and discussed in Chapter 1.

Next let us give a physical interpretation to the derivative of the delta function. By definition, we have

$$\langle \delta'(x), g(x) \rangle = -\langle \delta(x), g'(x) \rangle = -g'(0).$$

However, also by definition the derivative of the function $-g(x)$ evaluated at $x = 0$ is given as

$$-g'(0) = -\lim_{h \to 0} \frac{g(h) - g(0)}{h} = \lim_{h \to 0} \frac{g(0) - g(h)}{h}.$$

Now, using properties of the delta distribution, we find

$$-g'(0) = \lim_{h \to 0} \frac{\langle \delta(x), g(x) \rangle - \langle \delta(x - h), g(x) \rangle}{h}$$

$$= \lim_{h \to 0} \frac{\langle \delta(x) - \delta(x - h), g(x) \rangle}{h}.$$

Thus,

$$\delta'(x) = \lim_{h \to 0} \frac{\delta(x) - \delta(x - h)}{h}.$$

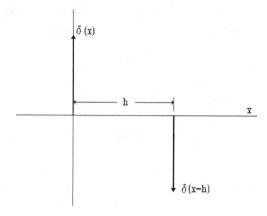

Figure 4.4   Physical interpretation of derivative of delta function.

The above equation does indeed conform to the definition of the derivative of $\delta(x)$ if it were considered to be a function.

Let us now consider the following physical interpretation. Shown in Figure 4.4 are two equal delta function of magnitude $1/h$ separated by the distance h. In the field of mechanics, if delta functions are used to represent concentrated forces, then the two of them taken together form a couple of unit magnitude. In the limit as $h \to 0$ we see that this couple reduces to a concentrated unit couple located at $x = 0$. In other words,

$$\lim_{h \to 0} h\delta'(x) = \text{unit couple located at } x = 0.$$

Earlier in this chapter we demonstrated that the derivative of a function at a jump discontinuity of magnitude b is given as a delta function of strength b located at the discontinuity. We now go over this again in a nonrigorous fashion for a function with a discontinuity of magnitude b located at $x = 0$. Consider the function shown in Figure 4.5a. This function is described mathematically as

$$f(x) = \begin{cases} \dfrac{xb}{h} & \text{for} \quad x \in [0,h], \\ b & \text{for} \quad x \in (h,\infty). \end{cases}$$

The derivative of this function (shown in Figure 4.5c) is given as

$$f'(x) = \begin{cases} \dfrac{b}{h} & \text{for} \quad x \in [0,h], \\ 0 & \text{for} \quad x \in (h,\infty). \end{cases}$$

As can be seen from Figure 4.5c, $f'(x)$ is a unit pulse function of height b/h and width h. We next consider both $f(x)$ and $f'(x)$ in the limit as $h \to 0$.

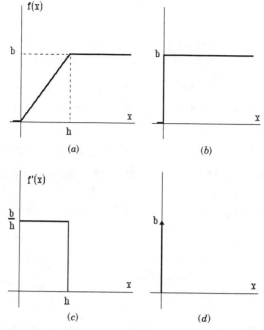

Figure 4.5 Derivative of jump discontinuity as limiting case.

As we see, in this limit $f'(x)$ goes to a delta function of strength b (Figure 4.5d) as was discussed in Chapter 1. Also, $f(x)$ goes to a function with a jump discontinuity of magnitude b at $x = 0$ (Figure 4.5b). From this demonstration we infer our desired results.

SUMMARY

In this chapter we have presented the concept of a distribution as a mapping of a particular class of functions $(g(x) \in S)$ to the set of complex numbers. We used the definition of the inner product as a guide to define various properties of distributions such as linearity, distributivity, and equality. We also showed that when the distribution in question was also a function, then the inner product integral was used to determine the complex number to which $g(x)$ was mapped. We also introduced several pure distributions (which were not necessarily functions) such as the delta, null, and comb. Finally we ended this chapter with a more intuitive discussion of the delta distribution. In this presentation we gave a physical interpretation to the results obtained previously using pure distribution theory.

PROBLEMS

1   Use the linearity property of the integral to demonstrate the multiplication by a constant property (A) of the inner product; that is, show

$$\langle af(x), g(x) \rangle = a \langle f(x), g(x) \rangle = \langle f(x), ag(x) \rangle.$$

2   Show that the inner product is distributive with respect to addition; that is, show

$$\langle af_1(x) + bf_2(x), g(x) \rangle = a \langle f_1(x), g(x) \rangle + b \langle f_2(x), g(x) \rangle.$$

3   Demonstrate that the inner product scale change property (F) requires $1/|a|$ and not simply $1/a$.

4   Determine if the following functions are elements of S, D, both, or neither.
   (a) $\sin(x)$
   (b) e^x for $x > 0$ or 0 otherwise
   (c) e^{-x} for $x > 0$ or 0 otherwise
   (d) $\ln(x)$
   (e) xe^{-x} for $x > 0$ or 0 otherwise.

5   Show that the sets S and D defined as part of the inner product are indeed vector spaces.

6   Show that if the first derivative of a function $g(x)$ exists on the interval $[a, b]$, then $g(x)$ is bounded over this interval.

7   To what value does the distribution (also a function) $f(x) = \sin^2(x)$ map the function $g(x) = 1/x^2$?

8   To what value does the distribution $f(x) = 1$ map the function $g(x) = \text{sinc}(2\pi x)$?

9   To what value does the distribution $\delta(x)(x - 3)^2$ map the function $g(x) = \cos(x)$?

10   What is the derivative of the comb distribution?

11   What sequence does the comb distribution of unit spacing (i.e., $\Delta x = 1$) map the function $g(x) = 1/x^2$ to?

12   If $t_1(x)$ maps $g(x) \in S$ to the value 7, what does $3t_1(x)$ map $g(x)$ to?

13   Consider the functions

$$f_1(x) = \begin{cases} x^2 \text{ for all } x \text{ not equal to an integer,} \\ -1 \text{ if } x \text{ is an integer.} \end{cases}$$

and

$$f_2(x) = \begin{cases} x^2 \text{ for all } x \text{ not equal to an integer,} \\ +1 \text{ if } x \text{ is an integer.} \end{cases}$$

If f_1 and f_2 are considered distributions, can they be considered equal? (Why)?

14 Show that if $t(x)$ is an odd distribution and $g(x)$ is an even function, then

$$\langle t(x), g(x) \rangle = 0.$$

15 If $t(x)$ is an even function and $g(x)$ is an odd function such that their inner product exist, then use the integral expression of equation (4.1) to show that it is equal to 0.

16 If $t_1(x)$ is a distribution that maps $g(x)$ to 7, and $t_2(x)$ is a distribution that maps $g(x)$ to 14, then what value will $4t_1(x) + 3t_2(x)$ map $g(x)$?

17 Suppose we know that $g(x)$ is a function that has a general form given as $g(x) = ax^2 + b$. Furthermore, assume that it is known that $\langle \delta(x), g(x) \rangle = 10$ and $\langle \delta(x - 1), g(x) \rangle = 15$. What are the numerical values of the constants a and b?

BIBLIOGRAPHY

Arsac, J., *Fourier Transforms and the Theory of Distributions*, Prentice-Hall, Englewood Cliffs, N.J., 1966.

Raven, F. H., *Mathematics of Engineering Systems*, McGraw-Hill, New York, N.Y., 1966.

5

THE FOURIER SERIES

In this chapter we discuss the Fourier series representation of a periodic function $f(t)$. In particular, we wish to find the required restrictions or conditions that the function $f(t)$ must satisfy in order to guarantee that the Fourier series coefficients exist and that the Fourier series representation will indeed converge to $f(t)$. To date, the search for both necessary and sufficient conditions that guarantee the convergence of the series to $f(t)$ has been unsuccessful.* However, using Lebesgue integration theory it is possible to state sufficient conditions in a rather clean and compact manner.

Obviously, one condition that the function must satisfy (by definition) is that it be periodic. Mathematically, we say a function $f(t)$ is periodic if there exists a constant $T > 0$ such that

$$f(t + T) = f(t) \tag{5.1}$$

for all t (and $t + T$) in the domain of definition of $f(t)$. The constant T is called the *period* of the function.

It should be clear that the sum, difference, product, and quotient of two functions of period T is also periodic. It should also be rather clear that if we know the behavior (i.e., can graph the function) on the interval $T = [a - T/2, a + T/2)$, then we know the behavior of the function on the entire real line because of its periodic nature. Note we have used the notation here that if a is any real number, then T denotes an interval of length T which extends from $a - T/2$ to $a + T/2$. Obviously the symbol "T" is doing double

---

*Necessary and sufficient conditions imply the series converges if, and only if, the conditions are met. Sufficient conditions imply the series converges if the conditions are met, although there may be other functions that do not satisfy the conditions and still its Fourier series converges.

duty here inasmuch as it is used to represent both the interval length and the interval itself. However, this should not lead to any real confusion because the context in which it is used will reveal its intended meaning.

At this point we note that for any periodic function we are able to shift the interval of integration by an arbitrary amount and not affect the value of the integral. In other words, when we integrate the function from $a + b - T/2$ to $a + b + T/2$ we obtain the same answer as integrating the function from $a - T/2$ to $a + T/2$. To show this we first note

$$\int_{a+b-T/2}^{a-T/2} f(t)\,dt = \int_{a+b-T/2}^{a-T/2} f(t+T)\,dt = \int_{a+b+T/2}^{a+T/2} f(\tau)\,d\tau$$

$$= -\int_{a+T/2}^{a+b+T/2} f(\tau)\,d\tau.$$

In the above equation we first made use of the fact that $f(t+T) = f(t)$ and then made the change of variable $\tau = t + T$ and modified the integration limits appropriately.

Now we write

$$\int_{a+b-T/2}^{a+b+T/2} f(t)\,dt = \int_{a+b-T/2}^{a-T/2} f(t)\,dt + \int_{a-T/2}^{a+T/2} f(t)\,dt + \int_{a+T/2}^{a+b+T/2} f(t)\,dt.$$

Now, using the previous result the first and last integral on the right-hand side of the above equation cancel and we obtain our desired result, that is,

$$\int_{a+b-T/2}^{a+b+T/2} f(t)\,dt = \int_{a-T/2}^{a+T/2} f(t)\,dt.$$

The trigonometric Fourier series representation of a function $f(t)$ over the interval $T = [a - T/2, a + T/2]$ is given as

$$f(t) = A_0 + \sum_{k=0}^{\infty} \left[A_k \cos\left(\frac{2\pi k t}{T}\right) + B_k \sin\left(\frac{2\pi k t}{T}\right) \right]. \tag{5.2}$$

The constants A_0, A_k, and B_k $(k = 1, 2, ..., \infty)$ are called the Fourier series coefficients of the function $f(t)$ and are given as

$$A_0 = \frac{1}{T} \int_T f(t)\,dt, \tag{5.3}$$

$$A_k = \frac{2}{T} \int_T f(t) \cos\left(\frac{2\pi k t}{T}\right) dt, \qquad k = 1, 2, ..., \tag{5.4}$$

$$B_k = \frac{2}{T} \int_T f(t) \sin\left(\frac{2\pi k t}{T}\right) dt, \qquad k = 1, 2, ... \tag{5.5}$$

The complex exponential form of the trigonometric Fourier series representation of a function $f(t)$ over the interval $T = [a - T/2, a + T/2]$ is given as

$$f(t) = \sum_{k=-\infty}^{\infty} C_k e^{2\pi i k t / T}. \tag{5.6}$$

The constants C_k ($k = -\infty, \ldots, -1, 0, 1, \ldots, \infty$) are called the *complex exponential Fourier series coefficients of the function* $f(t)$ and are given as

$$C_k = \left(\frac{1}{T}\right) \int_T f(t) e^{-2\pi i k t / T}, \qquad k \in (-\infty, \infty). \tag{5.7}$$

In this chapter we are particularly interested in the existence and convergence of the trigonometric Fourier series representation of a function as per equation (5.2). We approach this study along the line of Tolstov and first consider the broader situation of the Fourier series representation of a function by a general system of othogonal functions. In these general terms we consider two types of convergence. Namely, simply or ordinary convergence and convergence in the mean. Upon completion of this general study we limit our attention to the trigonometric Fourier series and present conditions that, when placed on a function $f(t)$, will guarantee convergence at both points of continuity and discontinuity. We then pay special attention to convergence at points of discontinuity and examine Gibbs' effect. Finally, in this chapter we present several theorems that simplify the calculation of the Fourier series coefficients and that can often be used to provide insight as to the convergence characteristics of the Fourier series.

GENERAL SYSTEM OF ORTHOGONAL FUNCTIONS

We begin by considering the following infinite system of real functions:

$$O_0(t), O_1(t), \ldots, O_n(t), \ldots. \tag{5.8}$$

This system is said to be an *orthogonal system* on the interval I if the following two conditions are satisfied:

$$\langle O_m(t), O_n(t) \rangle = \int_I O_m(t) O_n(t) \, dt = 0 \qquad \text{for } n \neq m, \tag{5.9}$$

$$\langle O_n(t), O_n(t) \rangle \neq 0. \tag{5.10}$$

Note that condition (5.10) implies that the integral

$$\int_I (O_n(t))^2 \, dt$$

exists for any n. Consequently, $O_n(t) \in L^2(I)$ for all n.

We define the *norm* of the function $O_n(t)$ as

$$\|O_n(t)\| = \langle O_n(t), O_n(t) \rangle^{1/2} = \left[\int_I (O_n(t))^2 \, dt\right]^{1/2}. \tag{5.11}$$

The system (5.8) is called *orthonormal* if conditions (5.9) and (5.10) are satisfied and if, *in addition,* $\|O_n(t)\| = 1$ for all n.

Let us now assume that we are able to find real constants c_n such that for some function $f(t)$ we have

$$f(t) = c_0 O_0(t) + c_1 O_1(t) + \cdots + c_n O_n(t) + \cdots. \tag{5.12}$$

At this point the equality of equation (5.12) is strictly conjecture. As a matter of fact, establishing conditions on the function $f(t)$ such that this equality is true is what this chapter is all about.

If equation (5.12) is indeed true, then we are able to demonstrate that

$$c_n = \frac{\langle f(t), O_n(t) \rangle}{\langle O_n(t), O_n(t) \rangle}. \tag{5.13}$$

The constants c_n are called the *Fourier series coefficients* with respect to the system of orthogonal functions (5.8). To demonstrate this we proceed by multiplying both sides of equation (5.12) by $O_n(t)$ and then integrating the results over the interval I, that is,

$$\int_I f(t) O_n(t) \, dt = c_0 \int_I O_0(t) O_n(t) + c_1 \int_I O_1(t) O_n(t) \, dt$$

$$+ \cdots + c_n \int_I O_n(t) O_n(t) \, dt + \cdots.$$

However, by equations (5.9) and (5.10) we arrive at

$$\int_I f(t) O_n(t) \, dt = c_n \int_I (O_n(t))^2 \, dt, \tag{5.14}$$

which is our desired result.

Again we stress the fact that the validity of the equality of equation (5.12) was necessary to show the above result. In other words, when equation (5.12) is true we establish equality of the integrals in equation (5.14), and since by assumption the integral on the right-hand side exists (see equation 5.10) we automatically establish the existence of the integral on the left. That is to say, the integral

$$\int_I f(t) O_n(t) \, dt$$

exists. Based on the presentations in Chapter 3 we also know that the above integral will exist if the product $f(t) O_n(t) \in L(I)$. Also, existence is guaranteed if $f \in L^2(I)$ since $O_n \in L^2(I)$, and by Theorem 3.22 this implies $f(t) O_n(t) \in L^2(I)$.

MINIMUM MEAN SQUARE ERROR

By conjecture, equality of equation (5.12) is valid for an infinite number of terms. In this section we consider the case when only a finite number of terms are used. In other words, we approximate the function $f(t)$ with the following finite series:

$$f(t) \sim C_0 O_0(t) + C_1 O_1(t) + \cdots + C_n O_n(t). \qquad (5.15)$$

For convenience we denote the right-hand side of equation (5.15) as $s_n(t)$, which is called the *nth partial sum* of the series. When this notation is used, equation (5.15) becomes

$$f(t) \sim s_n(t), \qquad (5.16)$$

where

$$s_n(t) = C_0 O_0(t) + C_1 O_1(t) + \cdots + C_n O_n(t). \qquad (5.17)$$

What we now wish to determine is "how well" $s_n(t)$ approximates $f(t)$. To answer this question we must first decide on some type of criteria which will measure the "goodness" of the fit. One of the most popular criteria is the minimum mean square error which is also called the "best least-squares fit." We begin by considering the quantity

$$\delta_n = \int_I (f(t) - s_n(t))^2 \, dt, \qquad (5.18)$$

which is called the *mean square error* in approximating $f(t)$ by $s_n(t)$. δ_n is the integral of the square of the difference between $f(t)$ and $s_n(t)$ over the interval I.

We now assume that $f \in L^2(I)$ and since $s_n(t) \in L^2(I)$ we must have $(f(t) - s_n(t)) \in L^2(I)$ (see Theorem 3.22). We now wish to choose the coefficients C_k in such a way as to minimize δ_n. We begin by expanding the square expression in equation (5.18), that is,

$$\delta_n = \int_I f^2(t)\,dt - 2\int_I f(t)s_n(t)\,dt + \int_I s_n^2(t)\,dt. \qquad (5.19)$$

Again, since both $f(t)$ and $s_n(t)$ are square summable, all the integrals in the above equation exist. Let us first consider the second term on the right-hand side of the above equation, that is,

$$2\int_I f(t)s_n(t)\,dt = 2\int_I f(t)C_0 O_0(t)\,dt + \cdots + 2\int_I f(t)C_n O_n(t)\,dt.$$

However, equation (5.13) implies

$$2\int_I f(t)s_n(t)\,dt = 2\sum_{k=0}^n C_k c_k \|O_k\|^2. \qquad (5.20)$$

Now let us consider the last term on the right-hand side of equation (5.19):

$$\int_I (s_n(t))^2\,dt = \int_I (C_0 O_0(t) + C_1 O_1(t) + \cdots + C_n O_n(t))^2\,dt.$$

By writing out a few terms in the above equation and applying the orthogonality relations of equation (5.13), we find

$$\int_I (s_n(t))^2 \, dt = C_0^2 \|O_0\|^2 + C_1^2 \|O_1\|^2 + \cdots + C_n^2 \|O_n\|^2. \tag{5.21}$$

Substitution of equations (5.20) and (5.21) into equation (5.19) yields

$$\delta_n = \int_I f^2(t) \, dt - 2 \sum_{k=0}^n C_k c_k \|O_k\|^2 + \sum_{k=0}^n C_k^2 \|O_k\|^2. \tag{5.22}$$

Finally, adding and subtracting $\sum c_k^2 \|O_k\|^2$ to the right-hand side of the above equation (5.22), along with a little rearrangement, yields

$$\delta_n = \int_I f^2(t) \, dt - \sum_{k=0}^n c_k^2 \|O_k\|^2 + \sum_{k=0}^n (C_k - c_k)^2 \|O_k\|^2. \tag{5.23}$$

We now wish to minimize δ_n with respect to the constants C_k. Obviously, the first two terms on the right-hand side of equation (5.23) are not a function of these constants. The last term is always positive and minimal when $C_k = c_k$. Thus, we see the minimum mean square error (or the best least squares fit) is obtained when the Fourier series coefficients are used as the constants in equation (5.17).

If we now return to equation (5.23) and consider this minimal condition (i.e., $C_k = c_k$), we obtain

$$D_n = \int_I f^2(t) \, dt - \sum_{k=0}^n c_k^2 \|O_k\|^2, \tag{5.24}$$

where D_n obviously denotes the minimum value of the mean square error.

Equation (5.24) reveals that as n increases, the nonnegative quantity D_n can only decrease. In other words, as n increases, the partial sums of the Fourier series $s_n(t)$ provide a better approximation to the function $f(t)$ (in terms of the mean square error).

BESSEL'S INEQUALITY AND CONVERGENCE IN THE MEAN

Since D_n in equation (5.24) is always greater than or equal to zero, it follows that

$$\int_I f^2(t) \, dt \geq \sum_{k=0}^n c_k^2 \|O_k\|^2. \tag{5.25}$$

Equation (5.25) is known as *Bessel's inequality*. We now note that since $f \in L^2(I)$, the integral expression on the left is bounded and consequently the increasing series

$$\sum_{k=0}^n c_k^2 \|O_k\|^2$$

is also bounded. Therefore it must converge (recall from Chapter 2 that a bounded increasing series must converge). This fact leads to the conclusion that

$$\lim_{n \to \infty} c_n \|O_n\| = 0. \tag{5.26}$$

For an orthonormal system, $\|O_n\|^2 = 1$ for all n and equations (5.25) and (5.26) become

$$\int_I f^2(t) \, dt \geq \sum_{k=0}^{n} c_k^2 \qquad \text{(orthonormal system)}, \tag{5.27}$$

$$\lim_{n \to \infty} c_n = 0 \qquad \text{(orthonormal system)}. \tag{5.28}$$

We say that the system (5.8) is *complete* if for any function $f \in L^2(I)$ the following equality is valid:

$$\int_I f^2(t) \, dt = \sum_{k=0}^{\infty} c_k^2 \|O_k\|^2, \tag{5.29}$$

where the c_k terms are the Fourier series coefficients of the function $f(t)$ as per equation (5.13). Equation (5.29) is known as *Parseval's equation* and/or the *completeness condition for system (5.8)*.

Consider the following limit:

$$\lim_{n \to \infty} \int_I [f(t) - s_n(t)]^2 \, dt = 0, \tag{5.30}$$

where $f \in L^2(I)$ and where $s_n(t)$ is given by equation (5.17) using the Fourier series coefficients as per equation (5.13). When the above limit of equation (5.30) is valid and equal to zero, we say the Fourier series *converges to f(t) in the mean*.

Combining the concept of a complete system and convergence in the mean, we are able to prove the following:

Theorem 5.1. A necessary and sufficient condition for the orthogonal system (5.8) to be complete is that the Fourier series of any square summable function $f(t)$ converges to $f(t)$ in the mean.

Proof. We have previously demonstrated (see equations (5.18), (5.23), and (5.24)) that for the Fourier series representation of a function the following equality is valid:

$$\int_I f^2(t) \, dt - \sum_{k=0}^{n} c_k^2 \|O_k\|^2 = \int_I [f(t) - s_n(t)]^2 \, dt. \tag{5.31}$$

Our desired results are directly obtained using this above equation. Let us first assume that system (5.8) is complete, that is,

$$\int_I f^2(t)\,dt = \sum_{k=0}^{\infty} c_k^2 \|O_k\|^2,$$

or, equivalently,

$$\lim_{n\to\infty}\left[\int_I f^2(t)\,dt - \sum_{k=0}^{n} c_k^2 \|O_k\|^2\right] = 0.$$

Thus, using equation (5.31) we have

$$\lim_{n\to\infty}\int_I [f(t) - s_n(t)]^2\,dt = 0,$$

which implies convergence in the mean and completes the sufficient condition portion of the proof. Next let us assume convergence in the mean, that is,

$$\lim_{n\to\infty}\int_I [f(t) - s_n(t)]^2\,dt = 0.$$

Again using equation (5.31) we see

$$\lim_{n\to\infty}\left[\int_I f^2(t)\,dt - \sum_{k=0}^{n} c_k^2 \|O_k\|^2\right] = 0,$$

or

$$\int_I f^2(t)\,dt = \sum_{k=0}^{\infty} c_k^2 \|O_k\|^2,$$

which implies completeness of the system and proves the necessary portion of the proof.   Q.E.D.

It is important to note that convergence in the mean of the Fourier series is different from ordinary or simple convergence of the Fourier series, which requires

$$\lim_{n\to\infty}\int_I [f(t) - s_n(t)]\,dt = 0.$$

More importantly, convergence in the mean does not necessarily imply ordinary convergence. In other words, completeness of the system does not necessarily guarantee ordinary convergence of the Fourier series to the function $f(t)$.

PROPERTIES OF COMPLETE SYSTEMS

In this section we consider several properties and consequences of complete systems. We first consider uniqueness and show that a Fourier series can converge in the mean to only one function.

Theorem 5.2 (Uniqueness). If the Fourier series of the function $f(t)$ converges in the mean to $f(t)$ and also to $g(t)$, then $f(t) = g(t)$ a.e. on I.

Proof. We assume that the Fourier series of the function $f(t)$ converges in the mean to both $f(t)$ and another function $g(t)$. We then obtain our desired result by showing that $f(t) = g(t)$ a.e. on I. By assumption,

$$\lim_{n\to\infty} \int_I [f(t) - s_n(t)]^2 \, dt = 0 \quad \text{and} \quad \lim_{n\to\infty} \int_I [g(t) - s_n(t)]^2 \, dt = 0.$$

Now consider

$$0 \le \int_I [f(t) - g(t)]^2$$

$$\le \int_I [f(t) - s_n(t) + s_n(t) - g(t)]^2 \, dt$$

$$\le 2 \int_I [f(t) - s_n(t)]^2 \, dt + 2 \int_I [g(t) - s_n(t)]^2 \, dt.$$

Note that in the above equation we have used the basic inequality

$$(f(t) + g(t))^2 \le 2(f^2(t) + g^2(t)).$$

Now in the limit as $n \to \infty$ we obtain

$$\int_I [f(t) - g(t)]^2 \, dt = 0,$$

which implies $f(t) = g(t)$ a.e. on I.   Q.E.D.

Note that in the above development we have taken the integral in the Lebesgue sense; thus, $f(t)$ and $g(t)$ are equal almost everywhere on the interval I. In other words, they may differ only on a set of measure zero or, equivalently, at a finite number of points.

Theorem 5.3. If the system (5.8) is complete, then any function $f(t)$ that is orthogonal to all functions of the system must be zero a.e. on I.

Proof. Since $f(t)$ is orthogonal to $O_n(t)$ $(n = 0, 1, ...)$, by definition we must have

$$\int_I f(t) O_n(t) \, dt = 0, \qquad n = 0, 1, \dots .$$

However, by equation (5.13) this implies that all the Fourier series coefficients are zero, that is $c_k = 0$, for $k = 0, 1, \dots$. Finally, the completeness condition (5.29) implies

$$\int_I f^2(t) \, dt = 0.$$

Consequently, $f(t) = 0$ a.e. on I.   Q.E.D.

Theorem 5.4. If the system (5.8) is complete, then the Fourier series of every square integrable function $f(t)$ can be integrated term by term, that is, if

$$f(t) = c_0 O_0(t) + c_1 O_1(t) + \cdots + c_n O_n(t) + \cdots,$$

then

$$\int_a^b f(t)\,dt = c_0 \int_a^b O_0(t)\,dt + c_1 \int_a^b O_1(t)\,dt + \cdots + c_n \int_a^b O_n(t)\,dt + \cdots,$$

where a and b are any two points in the interval such that $a < b$.

Proof. We begin by considering

$$\left| \int_a^b f(t)\,dt - \sum_{k=0}^n c_k \int_a^b O_k(t)\,dt \right| \le \int_a^b \left| f(t) - \sum_{k=0}^n c_k O_k(t) \right| dt$$

$$\le \int_I \left| f(t) - \sum_{k=0}^n c_k O_k(t) \right| dt.$$

Now using the Schwarz inequality (Theorem 3.21) we have

$$\le \left[\int_I (f(t) - \sum_{k=0}^n c_k O_k(t))^2\,dt \int_I 1\,dt \right]^{1/2}.$$

However, the completeness condition implies that in the limit as $n \to \infty$ the first integral in the above expression approaches zero. Thus we obtain

$$\int_a^b f(t)\,dt = \lim_{n \to \infty} \sum_{k=0}^n c_k \int_a^b O_k(t)\,dt,$$

which is our desired result.   Q.E.D.

We note that the results of Theorem 5.4 are true whether or not the Fourier series converges in the ordinary sense.

TRIGONOMETRIC SYSTEM OF ORTHOGONAL FUNCTIONS

We now limit our attention to the system of trigonometric functions and the interval $T = [a - T/2, a + T/2]$. In this situation the system (5.8) is given specifically as

$$O_0(t) = 1,$$

$$O_{2k}(t) = \cos\left(\frac{2\pi k t}{T}\right), \tag{5.32}$$

$$O_{2k-1}(t) = \sin\left(\frac{2\pi k t}{T}\right).$$

We now demonstrate that system (5.32) is orthogonal. To accomplish this we make use of the following basic trigonometric identities:

$$\cos(\alpha + \beta) = \cos(\alpha)\cos(\beta) - \sin(\alpha)\sin(\beta), \qquad (5.33)$$

$$\sin(\alpha + \beta) = \sin(\alpha)\cos(\beta) + \sin(\beta)\cos(\alpha). \qquad (5.34)$$

Elementary manipulation of the above the two equations yields

$$\cos(\alpha)\cos(\beta) = \tfrac{1}{2}[\cos(\alpha + \beta) + \cos(\alpha - \beta)], \qquad (5.35)$$

$$\sin(\alpha)\sin(\beta) = \tfrac{1}{2}[\cos(\alpha - \beta) - \cos(\alpha + \beta)], \qquad (5.36)$$

$$\sin(\alpha)\cos(\beta) = \tfrac{1}{2}[\sin(\alpha + \beta) + \sin(\alpha - \beta)]. \qquad (5.37)$$

Obviously the functions in system (5.32) are periodic with period T. We now demonstrate that for any integer n, we have

$$\int_T \cos\left(\frac{2\pi nt}{T}\right) dt = 0 \qquad (5.38)$$

and

$$\int_T \sin\left(\frac{2\pi nt}{T}\right) dt = 0. \qquad (5.39)$$

At the beginning of this chapter we showed that for a periodic function we were always able to shift the interval of integration by an arbitrary amount and not effect the value of the integral. Consequently, we demonstrate the results of equations (5.38) and (5.39) over the simple interval $T = [-T/2, T/2]$. We begin with equation (5.38):

$$\int_{-T/2}^{T/2} \cos\left(\frac{2\pi nt}{T}\right) dt = \left(-\frac{T}{2\pi n}\right) \sin\left(\frac{2\pi nt}{T}\right)\Bigg|_{-T/2}^{T/2}$$

$$= \left(-\frac{T}{2\pi n}\right)[\sin(\pi n) + \sin(\pi n)] = 0.$$

The above result is true for *all* integers $n \neq 0$. Now we examine equation (5.39):

$$\int_{-T/2}^{T/2} \sin\left(\frac{2\pi nt}{T}\right) dt = \left(-\frac{T}{2\pi n}\right) \cos\left(\frac{2\pi nt}{T}\right)\Bigg|_{-T/2}^{T/2}$$

$$= \left(-\frac{T}{2\pi n}\right)[\cos(\pi n) - \cos(\pi n)] = 0.$$

The above result is true for *all* integers. In particular, the result is also true when $n = 0$ as can be appreciated by applying L'Hôpital's rule.

We next demonstrate the fact that the trigonometric system (5.32) is orthogonal. Because of the way in which this system was defined, we have several situations that we must examine. First we consider three specific cases:

$$\langle O_0(t), O_{2n-1}(t) \rangle = \int_T \sin\left(\frac{2\pi nt}{T}\right) dt = 0,$$

$$\langle O_0(t), O_{2n}(t) \rangle = \int_T \cos\left(\frac{2\pi nt}{T}\right) dt = 0,$$

$$\langle O_0(t) O_0(t) \rangle = \int_T dt = T.$$

We now consider the more general situations. First, $\langle O_m(t), O_n(t) \rangle$, where $m \neq n$ and both m and n are even, that is,

$$\langle O_m(t), O_n(t) \rangle = \int_T \cos\left(\frac{2\pi mt}{T}\right) \cos\left(\frac{2\pi nt}{T}\right) dt.$$

Using equation (5.35) we rewrite the above equation as

$$\langle O_m(t), O_n(t) \rangle = \frac{1}{2} \int_T \cos\left(\frac{2\pi(n+m)t}{T}\right) dt$$

$$+ \frac{1}{2} \int_T \cos\left(\frac{2\pi(n-m)t}{T}\right) dt.$$

However, since $n + m$ and $n - m$ are integer values not equal to zero, equation (5.38) tells us that both of the above integrals are zero, Thus,

$$\langle O_m(t), O_n(t) \rangle = 0 \quad \text{for} \quad n \neq m \quad (n \text{ and } m \text{ even integers}).$$

Next we consider the situation when both m and n are even integers and $n = m$. Using the same logic as above we again arrive at

$$\langle O_n(t), O_n(t) \rangle = \frac{1}{2} \int_T \cos\left(\frac{2\pi(n+n)t}{T}\right) dt$$

$$+ \frac{1}{2} \int_T \cos\left(\frac{2\pi(n-n)t}{T}\right) dt.$$

Using the same reasoning as before we see that the first integral in the above equation is zero. However, since $n = m$, the second integral remains and the above equation becomes

$$\langle O_m(t), O_n(t) \rangle = \frac{1}{2} \int_T dt = \frac{1}{2}(T) = \frac{T}{2}.$$

Similar reasoning shows that for both n and m odd we have

$$\langle O_m(t), O_n(t) \rangle = \int_T \sin\left(\frac{2\pi mt}{T}\right) \sin\left(\frac{2\pi nt}{T}\right) dt = 0$$

$$\text{for} \quad n \neq m \quad (m \text{ and } n \text{ odd})$$

$$\langle O_n(t), O_n(t) \rangle = \frac{T}{2}.$$

Finally, as our last situation, we examine the case when n is odd and m is even, that is,

$$\langle O_m(t), O_n(t) \rangle = \int_T \sin\left(\frac{2\pi mt}{T}\right) \cos\left(\frac{2\pi nt}{T}\right) dt.$$

Using equation (5.37) we rewrite the above equation as

$$\langle O_m(t), O_n(t) \rangle = \frac{1}{2} \int_T \sin\left(\frac{2\pi(n+m)t}{T}\right) dt + \frac{1}{2} \int_T \sin\left(\frac{2\pi(n-m)t}{T}\right) dt.$$

Using equation (5.39) we see that both of the above integrals are equal to zero. Thus,

$$\langle O_m(t), O_n(t) \rangle = 0.$$

We formally summarize the previous development as the following:

Theorem 5.5. The trigonometric system (5.32) is orthogonal.

Now that we have demonstrated that the trigonometric system (5.32) is orthogonal, the results of the previous sections can be directly applied to obtain the following results.

Theorem 5.6. If the function $f(t)$ can be represented by the infinite Fourier series

$$f(t) = A_0 + \sum_{k=1}^{\infty} A_k \cos\left(\frac{2\pi kt}{T}\right) + B_k \sin\left(\frac{2\pi kt}{T}\right),$$

then the Fourier series coefficients are given as per equations (5.3)–(5.5), that is,

$$A_0 = \frac{1}{T} \int_T f(t) dt, \tag{5.3}$$

$$A_k = \frac{2}{T} \int_T f(t) \cos\left(\frac{2\pi kt}{T}\right) dt, \tag{5.4}$$

$$B_k = \frac{2}{T} \int_T f(t) \sin\left(\frac{2\pi kt}{T}\right) dt. \tag{5.5}$$

Proof. The proof of this theorem follows by direct application of equation (5.13).   Q.E.D.

Theorem 5.7. If the square-summable function $f \in L^2(T)$ is approximated by the finite series

$$f(t) \sim a_0 + \sum_{k=1}^{n} a_k \cos\left(\frac{2\pi kt}{T}\right) + b_k \sin\left(\frac{2\pi kt}{T}\right),$$

then the best approximation in terms of the mean square error is obtained when the coefficients are obtained using equations (5.3)–(5.5), that is, when the coefficients are chosen to be the Fourier series coefficients of system (5.32).

Theorem 5.8 (Bessel's Inequality). For the trigonometric system (5.32), if $f \in L^2(T)$, then the following inequality is valid:

$$\frac{2}{T} \int_T f^2(t)\,dt \geq 2A_0^2 + \sum_{k=1}^{n}(A_k^2 + B_k^2).$$

Proof. The proof of this theorem follows directly from equation (5.25) and the fact that $\|O_0\|^2 = T$ and $\|O_k\|^2 = T/2$.   Q.E.D.

Theorem 5.9. If $f \in L^2(T)$, then

$$\lim_{n\to\infty} \int_T f(t)\cos\left(\frac{2\pi nt}{T}\right) dt = 0 \quad \text{and} \quad \lim_{n\to\infty} \int_T f(t)\sin\left(\frac{2\pi nt}{T}\right) dt = 0.$$

Proof. It follows directly from equation (5.26) that

$$\lim_{n\to\infty} A_n = 0 \quad \text{and} \quad \lim_{n\to\infty} B_n = 0.$$

However,

$$A_n = \frac{2}{T} \int_T f(t)\cos\left(\frac{2\pi nt}{T}\right) dt \quad \text{and} \quad B_n = \frac{2}{T} \int_T f(t)\sin\left(\frac{2\pi nt}{T}\right) dt,$$

from which we obtain our desired results.   Q.E.D.

We say the trigonometric Fourier series *converges in the mean* to the square-summable function $f(t)$ if the following limit is valid:

$$\lim_{n\to\infty} \int_T \left[f(t) - A_0 - \sum_{k=1}^{n} A_k \cos\left(\frac{2\pi kt}{T}\right) + B_k \sin\left(\frac{2\pi kt}{T}\right) \right]^2 dt = 0.$$

Combining the above definition with Theorem 5.1 and equation (5.30) we have the following theorem.

Theorem 5.10 (Parseval's Equation). If the trigonometric Fourier series converges in the mean to the square-summable function $f(t)$, then

$$\frac{2}{T} \int_T f^2(t)\,dt = 2A_0^2 + \sum_{k=1}^{\infty}(A_k^2 + B_k^2).$$

COMPLEX EXPONENTIAL SYSTEM OF ORTHOGONAL FUNCTIONS

In this section we consider the system of complex exponential functions given by

$$O_n(t) = e^{2\pi int/T}, \qquad n \in (-\infty, \infty). \tag{5.40}$$

We first note that the function $O_n(t)$ are periodic with period T, that is,

$$O_n(t + T) = e^{2\pi in(t+T)/T} = e^{2\pi int/T} e^{2\pi in} = e^{2\pi int/T}.$$

(Here we have used the fact that $\exp[2\pi in] = 1$ for any integer n).
We now demonstrate orthogonality for this system.

$$\langle O_n(t), O_m(t) \rangle = \int_T e^{2\pi i(n+m)t/T} \, dt$$

$$= \left(\frac{T}{2\pi i(n + m)} \right) [e^{i\pi(m+n)} - e^{-i\pi(n+m)}]$$

$$= \frac{T \sin(\pi(n + m))}{\pi(n + m)} = 0, \qquad n \neq m. \tag{5.41}$$

Note. In the above equation we used Euler's equations, that is,

$$\sin(x) = \frac{e^{ix} - e^{-ix}}{2i}, \qquad \cos(x) = \frac{e^{ix} + e^{-ix}}{2}.$$

Also we note

$$\langle O_{-n}(t), O_n(t) \rangle = \int_T e^{2\pi i(n-m)t/T} \, dt = \int_T dt = T.$$

Equations (5.41) and (5.42) are the orthogonality conditions for system (5.40). As we did earlier in this chapter, for the general system (5.8) let us assume that we are able to find real constants C_n such that for some function $f(t)$ we have

$$f(t) = \sum_{n=-\infty}^{\infty} C_n e^{2\pi int/T}. \tag{5.6}$$

If indeed equation (5.16) is valid, then we are able to show

$$C_n = \frac{1}{T} \int_T f(t) e^{-2\pi int/T} \, dt. \tag{5.7}$$

This is accomplished by multiplying both sides of equation (5.6) by $\exp[-2\pi int/T]$ and applying the orthogonality equations (5.41) and (5.42). Using reasoning similar to that presented in previous sections, we can also establish the concept of convergence in the mean, minimum mean square error, uniqueness, and so on. These developments are left as an exercise for the reader (see problems 8–12).

THE RIEMANN LEBESGUE LEMMA

In this section we present a result commonly referred to as the *Riemann Lebesgue lemma*, which is one of our most useful tools in deriving convergence conditions for the Fourier series.

Theorem 5.11 (Riemann Lebesgue). If $f \in L(T)$, then for every real numbers α and β we have

$$\lim_{\alpha \to \infty} \int_T f(t)\sin(\alpha t + \beta)\,dt = 0.$$

Proof. This result is obviously true if $f(t)$ is a constant (equal to A) over the interval T since

$$\left| \int_T A\sin(\alpha t + \beta)\,dt \right| = \left| \left(\frac{A}{\alpha} \right) \cos(\alpha t + \beta) \right|_{-T/2}^{T/2}$$

$$= \left| \left(\frac{A}{\alpha} \right) \left[\cos\left(\frac{\alpha T}{2} + \beta \right) - \cos\left(-\frac{\alpha T}{2} + \beta \right) \right] \right| \le 2\left| \frac{A}{\alpha} \right|.$$

Thus, in the limit as $\alpha \to \infty$ we see that

$$\int_T A\sin(\alpha t + \beta)\,dt = 0.$$

Now since this result is true for any constant function on T, it is also true for any step function defined on T since we can always divide the interval into a finite number of subintervals over which the function is a constant. Thus, for any step function $s(t)$ defined on T, we have: Given any $\varepsilon > 0$, there exists a positive number M such that $\alpha > M$ implies that

$$\left| \int_T s(t)\sin(\alpha t + \beta)\,dt \right| < \frac{\varepsilon}{2}.$$

However, since $f \in L(T)$ we know (by definition) that there exists a sequence of step functions $\{s_n(t)\}$ such that, given any $\varepsilon > 0$, there exists an integer N such that $n > N$ implies

$$\left| \int_T [f(t) - s_n(t)]\,dt \right| < \frac{\varepsilon}{2}.$$

Consequently,

$$\left| \int_T f(t)\sin(\alpha t + \beta)\,dt \right| = \left| \int_T [f(t) - s_n(t) + s_n(t)]\sin(\alpha t + \beta)\,dt \right|$$

$$\leq \left| \int_T [f(t) - s_n(t)]\sin(\alpha t + \beta)\,dt \right|$$

$$+ \left| \int_T s_n(t)\sin(\alpha t + \beta)\,dt \right|$$

$$\leq \left| \int_T [f(t) - s_n(t)]\,dt \right| + \left| \int_T s_n(t)\sin(\alpha t + \beta)\,dt \right|.$$

Thus, choosing $n > N$ and $\alpha > M$ we find

$$\left| \int_T f(t)\sin(\alpha t + \beta)\,dt \right| < \frac{\varepsilon}{2} + \frac{\varepsilon}{2} = \varepsilon,$$

and we have established our desired results. Q.E.D.

Recall that in a previous section (Theorem 5.9) we proved that if the function $f(t)$ was square summable (i.e., $f \in L^2(T)$), then the following limits were valid:

$$\lim_{n\to\infty} \int_T f(t)\cos\left(\frac{2\pi nt}{T}\right)\,dt = 0 \quad \text{and} \quad \lim_{n\to\infty} \int_T f(t)\cos\left(\frac{2\pi nt}{T}\right)\,dt = 0.$$

Theorem 5.11 tells us that these limits are also true when the function is Lebesgue integrable. In other words, if we let $\alpha = 2\pi n/T$ and $\beta = 0$ in Theorem 5.11, then we obtain the first limit. The second limit is obtained using $\alpha = 2\pi nt/T$ and $\beta = \pi/2$.

THE DIRICHLET KERNEL FUNCTION

In this section we present the Dirichlet kernel, which is crucial to our development of pointwise convergence of the Fourier series as well as Gibbs' phenomenon.

The *Dirichlet kernel function* is defined as follows:

$$D_n(x) = \frac{\sin([n + \frac{1}{2}]x)}{2\sin(x/2)}. \tag{5.43}$$

We now demonstrate that

$$D_n(x) = \tfrac{1}{2} + \cos(x) + \cos(2x) + \cdots + \cos(nx). \tag{5.44}$$

To obtain the above result we write

$$S = \tfrac{1}{2} + \cos(x) + \cos(2x) + \cdots + \cos(nx).$$

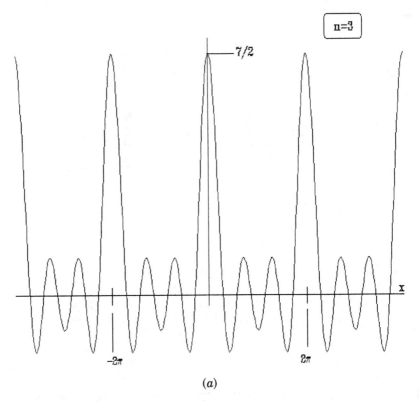

$n=3$

$7/2$

x

-2π

2π

(*a*)

Figure 5.1 The Dirichlet kernel function: (a) *n*= 3, (b) *n*= 6, (c) *n*= 12.

Next we multiply both sides of the above equation by $2\sin(x/2)$, that is,

$$2S\sin\left(\frac{x}{2}\right) = \sin\left(\frac{x}{2}\right) + 2\cos(x)\sin\left(\frac{x}{2}\right) + \cdots + 2\cos(nx)\sin\left(\frac{x}{2}\right).$$

Now using a slight rearrangement of equation (5.37) we find

$$2S\sin\left(\frac{x}{2}\right) = \sin\left(\frac{x}{2}\right) + \sin\left(x + \frac{x}{2}\right) - \sin\left(\frac{x}{2}\right) + \sin\left(2x + \frac{x}{2}\right)$$
$$- \sin\left(x + \frac{x}{2}\right) + \cdots + \sin\left(nx - \frac{x}{2}\right) + \sin\left(nx + \frac{x}{2}\right)$$
$$- \sin\left(nx - \frac{x}{2}\right).$$

In the above equation, every other term cancels and the resulting equation "telescopes" to

$$2S\sin\left(\frac{x}{2}\right) = \sin([n + \tfrac{1}{2}]x),$$

which leads directly to our desired result.

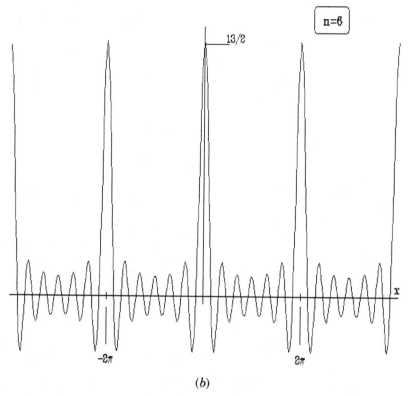

(b)

Figure 5.1 *(Continued)*

Shown in Figure 5.1 is a plot of the Dirichlet kernel function (from -4π to $+4\pi$) for $n = 3$, 6, and 12, respectively. This function is periodic with period 2π, as can be seen in the figure and which is demonstrated as follows:

$$D_n(x + 2\pi) = \frac{\sin([n + \frac{1}{2}][x + 2\pi])}{\sin([x + 2\pi]/2)}$$

$$= \frac{\sin([n + \frac{1}{2}]x + 2n\pi + \pi)}{\sin(x/2 + \pi)}$$

$$= \frac{\sin([n + \frac{1}{2}]x + \pi)}{\sin(x/2 + \pi)} = \frac{-\sin([n + \frac{1}{2}]x)}{-\sin(x/2)} = D_n(x).$$

The above result can also be demonstrated using equation (5.44) and noting that each cosine term in this expression is periodic with period 2π.

We should also note from equation (5.44) that $D_n(x)$ is an even function, that is,

$$D_n(-x) = D_n(x).$$

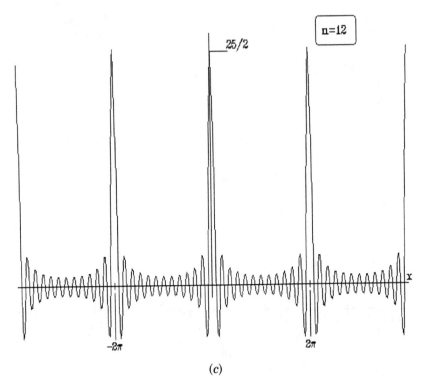

(c)

Figure 5.1   (Continued)

While the overall period of $D_n(x)$ is 2π, we note that there is also a subpe-
riodic higher frequency behavior of the function. This is obviously due to the
numerator sine function (see equation (5.43)), which has a basic frequency of

$$w_D = \frac{2n+1}{4\pi},$$

or, in other words, a basic period of

$$T_D = \frac{1}{w_D} = \frac{4\pi}{2n+1}.$$

Clearly, this results in $n + \frac{1}{2}$ cycles of $D_n(x)$ per 2π period. This higher frequency content of $D_n(x)$ can be directly related to the frequency $(n/2\pi)$ of
the last cosine term used in the summation of equation (5.44).

The Dirichlet kernel function takes on its global maximum values when
$x = 2\pi k$ ($k = -\infty, \ldots, -1, 0, 1, \ldots, \infty$). In particular, when $k = 0$ we determine
(using L'Hôpital's rule) that this maximum value is equal to $(2n + 1)/2$. Finally,
we can show

$$\int_0^{2\pi} D_n(x)\,dx = \pi. \tag{5.45}$$

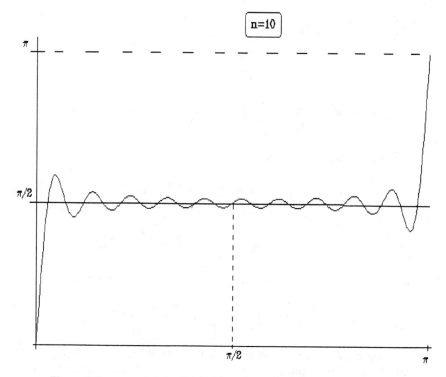

Figure 5.2   Integral of the Dirichlet kernel as a function of x from 0 to 2π.

This follows directly using term-by-term integration in equation (5.44) and noting that the integral of each cosine function vanishes over the interval $[0, 2\pi]$. In Figure 5.2 we show the integral of the Dirichlet kernel function as an antiderivative, or function of x, from 0 to 2π.

POINTWISE CONVERGENCE OF THE FOURIER SERIES

We now consider the pointwise convergence of the Fourier series. That is to say, we examine the convergence of the Fourier series to the function $f(t)$ for any point t in the interval T. To accomplish this, we first derive an integral expression for the nth partial sum of the Fourier series. We begin with the expression for the nth partial sum of the Fourier series approximation to the function $f(t)$, that is,

$$s_n(t) = A_0 + \sum_{k=1}^{n} A_k \cos\left(\frac{2\pi k t}{T}\right) + B_k \sin\left(\frac{2\pi k t}{T}\right). \qquad (5.46)$$

We now substitute equations (5.3)–(5.5) for the Fourier series coefficients A_0, A_k, and B_k into equation (5.56) to obtain

$$s_n(t) = \frac{1}{T} \int_T f(x)dx + \sum_{k=1}^{n} \left(\frac{2}{T}\right) \int_T f(x)\cos\left(\frac{2\pi k x}{T}\right)\cos\left(\frac{2\pi k t}{T}\right) dt$$

$$+ \sum_{k=1}^{n} \left(\frac{2}{T}\right) \int_T f(x)\sin\left(\frac{2\pi k x}{T}\right)\sin\left(\frac{2\pi k t}{T}\right) dt.$$

Factoring and regrouping the above equation, we find

$$s_n(t) = \frac{1}{T} \int_T f(x)\left[1 + \sum_{k=1}^{n} 2\cos\left(\frac{2\pi k x}{T}\right)\cos\left(\frac{2\pi k t}{T}\right)\right.$$

$$\left. + 2\sin\left(\frac{2\pi k x}{T}\right)\sin\left(\frac{2\pi k t}{T}\right)\right].$$

Now using the trigonometric identity of equation (5.33), we obtain

$$s_n(t) = \frac{2}{T} \int_T f(x)\left[\frac{1}{2} + \sum_{k=1}^{n} \cos\left(\frac{2\pi k(x-t)}{T}\right)\right] dx.$$

Finally, using equation (5.44) we are able to write the above equation for the nth partial sum of the Fourier series in terms of the Dirichlet kernel function, that is,

$$s_n(t) = \frac{2}{T} \int_T f(x)D_n\left(\frac{2\pi(x-t)}{T}\right) dx.$$

We now take advantage of the periodic nature of both $f(x)$ and $D_n(x)$ to rewrite equation (5.47). We begin by splitting the integral, that is,

$$s_n(t) = \frac{2}{T} \int_{-T/2}^{0} f(x)D_n\left(\frac{2\pi(x-t)}{T}\right) dx + \frac{2}{T} \int_{0}^{T/2} f(x)D_n\left(\frac{2\pi(x-t)}{T}\right) dx.$$

We now make the substitution of variables $x - t = u$ or $x = u + t$ and $dx = du$. This results in

$$s_n(t) = \left(\frac{2}{T}\right) \int_{-t-T/2}^{-t} f(u+t)D_n\left(\frac{2\pi u}{T}\right) du$$

$$+ \frac{2}{T} \int_{-t}^{-t+T/2} f(u+t)D_n\left(\frac{2\pi u}{T}\right) du.$$

Since both f and D_n are periodic we can shift the interval of integration by an amount t without affecting the result, that is,

$$s_n(t) = \frac{2}{T} \int_{-T/2}^{0} f(u+t)D_n\left(\frac{2\pi u}{T}\right) du + \frac{2}{T} \int_{0}^{T/2} f(u+t)D_n\left(\frac{2\pi u}{T}\right) du.$$

Finally, in the first integral we let $u = -u$ (and $du = -du$), change and invert the limits of integration, and note the fact that $D_n(u)$ is an even function. This leads to the following:

$$s_n(t) = \frac{2}{T} \int_0^{T/2} [f(t+u) + f(t-u)] D_n \left(\frac{2\pi u}{T} \right) du. \qquad (5.48)$$

Convergence of the Fourier series at any point $t \in T$ is determined by the behavior of $s_n(t)$ in the limit as $n \to \infty$, and equation (5.48) allows us to study convergence of the series by considering the following integral limit:

$$\lim_{n \to \infty} \left(\frac{2}{T} \right) \int_0^{T/2} [f(t+u) + f(t-u)] D_n \left(\frac{2\pi u}{T} \right) du.$$

We note that the integral in the above equation is performed using the variable u over the interval 0 to $T/2$. In terms of the function $f(t)$, we are integrating about the point t a distance $T/2$ in both directions (i.e., $f(t+u)$ and $f(t-u)$). We now present the Riemann Localization Theorem, which says that (in the limit as $n \to \infty$) we need only evaluate this integral over an arbitrarily small neighborhood of radius δ about the point t.

Theorem 5.12 (Riemann Localization Theorem). For a Lebesgue integrable function $f(t)$, the behavior of the Fourier series at a point $t \in T$ depends only on the values of $f(t)$ in an arbitrarily small neighborhood of the point t.

Proof. Assume $f \in L(T)$ and define the function

$$F(t) = \frac{f(t+u) + f(t-u)}{2\sin(\pi u/T)}.$$

Since $f \in L(T)$ and $1/(2\sin(\pi u/T))$ is continuous and bounded for $u > 0$ we have $F(t) \in L(\delta, T/2)$, where δ is an arbitrarily small positive real number.

We now use equation (5.48) with the Dirichlet kernel written as per equation (5.43), that is,

$$s_n(t) = \frac{2}{T} \int_0^{T/2} \left[\frac{(f(t+u) + f(t-u))}{2\sin(\pi u/T)} \right] \sin \left[\frac{2\pi(n+\frac{1}{2})u}{T} \right] du.$$

or

$$s_n(t) = \left(\frac{2}{T} \right) \int_0^{T/2} F(t) \sin \left[\frac{2\pi(n+\frac{1}{2})u}{T} \right] du.$$

We now split this integral as follows:

$$s_n(t) = \frac{2}{T} \int_0^{\delta} F(t) \sin \left[\frac{2\pi(n+\frac{1}{2})u}{T} \right] du + \frac{2}{T} \int_{\delta}^{T/2} F(t) \sin \left[\frac{2\pi(n+\frac{1}{2})u}{T} \right] du.$$

Now since $F(t) \in L(\delta, T/2)$ we use the Riemann Lebesgue Theorem (Theorem 5.11) to show that in the limit as $n \to \infty$ the second integral in the above equation vanishes and we are left with

$$\lim_{n \to \infty} s_n(t) = \lim_{n \to \infty} \left(\frac{2}{T} \right) \int_0^\delta [f(t+u) + f(t-u)] D_n \left(\frac{2\pi u}{T} \right) du,$$

which is our desired result.   Q.E.D.

Theorem 5.12 tells us that the convergence (or divergence) of a Fourier series, at a particular point t, is governed entirely by the behavior of the function in an arbitrarily small neighborhood of t. This is rather interesting inasmuch as the Fourier series coefficients depend upon the values which the function $f(t)$ assumes throughout the entire interval T (see equations (5.3)–(5.5)). It is also very interesting to note that this theorem proves that if the Fourier series of $f(t)$ is originally convergent at t, then no matter how we change the values of the function (while leaving it Lebesgue integrable) outside an arbitrarily small neighborhood of t, the series remains convergent at t.

Given a Lebesgue integrable function $f(t)$, let us assume that at any point $t \in T$ the following limits exist:

$$\lim_{h \to 0} f(t+h) = f(t^+) \quad \text{and} \quad \lim_{h \to 0} f(t-h) = f(t^-).$$

Also, let $s(t)$ denote the limit as $n \to \infty$ of the nth partial sum of the Fourier series, that is,

$$s(t) = \lim_{n \to \infty} s_n(t).$$

We now examine the relationship between $s(t)$ and $f(t)$ at a point of continuity and at a point of discontinuity. We begin by using the results of the Riemann Localization Theorem (Theorem 5.12) and writing

$$\lim_{n \to \infty} \left(\frac{2}{T} \right) \int_0^\delta \frac{[f(t+u) + f(t-u)]}{2\sin(\pi u/T)} \sin \left(\frac{(2n+1)\pi u}{T} \right) du. \tag{5.49}$$

Now, since δ can be made arbitrarily small (i.e., approach zero in the limit) we can assume that if $u \in (0, \delta)$, then

$$f(t+u) + f(t-u) = f(t^+) + f(t^-),$$

which may be considered constant with respect to u. Consequently, it can be removed from under the integral sign in equation (5.49). Furthermore, for $u \in (0, \delta)$ we have

$$2\sin \left(\frac{\pi u}{T} \right) \sim \frac{2\pi u}{T}.$$

When these considerations are used in equation (5.49) we obtain

$$s(t) = \frac{f(t^+) + f(t^-)}{\pi} \lim_{n \to \infty} \int_0^\delta \frac{\sin(2n+1)\pi u/T}{u} du. \tag{5.50}$$

Finally, let us make the change of variable $\tau = (2n + 1)\pi u/T$ or

$$u = \frac{\tau T}{(2n + 1)\pi} \quad \text{and} \quad du = \frac{T d\tau}{(2n + 1)\pi}.$$

When this is used in equation (5.50) we obtain

$$s(t) = \frac{f(t^+) + f(t^-)}{\pi} \lim_{n \to \infty} \int_0^{(2n+1)\pi\delta/T} \frac{\sin(\tau)}{\tau} d\tau.$$

However, as is well known, the above integral is equal to $\pi/2$. Thus,

$$s(t) = \frac{f(t^+) + f(t^-)}{2}. \tag{5.51}$$

Let us first assume that $f \in L(T)$ is continuous at the point t. Thus, $f(t) = f(t^+) = f(t^-)$. Consequently, by equation (5.51) we have $s(t) = f(t)$. Thus, we see that when the function is continuous at the point t, the Fourier series representation of that function converges to $f(t)$.

Now let us assume that $f \in L(T)$ has a jump discontinuity at the point t. In this case, $f(t^-)$ is the value of the function just before the jump, and $f(t^+)$ is the value of the function just after the jump. In this case, equation (5.51) implies $s(t) = [f(t^+) + f(t^-)]/2$, or, in other words, the Fourier series converges to the average of the values of the function on either side of the jump.

GIBBS' PHENOMENON

In the previous section we examined the convergence of the Fourier series at a point in the interval T. This situation was studied as the number of terms in series approached infinity. In this section we consider the convergence of the Fourier series at a point of discontinuity when only a finite number of terms are used to represent the function. This situation leads to the familiar *Gibbs' effect* or ringing about the discontinuity.

There are many plausible ways in which a function $f(t)$ can possess a discontinuity. However, since we wish to illustrate basic principles and the behavior of the Fourier series at a discontinuity, we choose a rather simple jump discontinuity located at $t = 0$. Specifically, let us consider the function shown in Figure 5.3, which is mathematically defined as

$$f(t) = \begin{cases} \frac{1}{2}, & t \geq 0, \\ -\frac{1}{2}, & t < 0. \end{cases} \tag{5.52}$$

Returning to equation (5.48), we see that the nth partial sum of the series at any point $t \in T$ is given as

$$s_n(t) = \frac{2}{T} \int_0^{T/2} [f(t + x) + f(t - x)] D_n\left(\frac{2\pi x}{T}\right) dx. \tag{5.53}$$

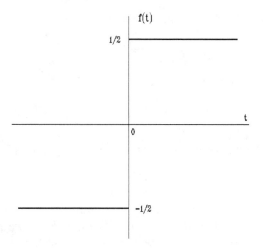

Figure 5.3   Function with unity magnitude jump discontinuity at $t = 0$.

Let us now consider the behavior of the expression $(f(t + x) + f(t - x))$ as a function of x when $f(t)$ is given as per equation (5.52). For convenience* we only consider values of $t \geq 0$:

If $x \leq t$, then $f(t + x) + f(t - x) = f(+) + f(+) = \frac{1}{2} + \frac{1}{2} = 1$.

If $x > t$, then $f(t + x) + f(t - x) = f(+) + f(-) = \frac{1}{2} - \frac{1}{2} = 0$.

When these considerations are taken into account, equation (5.53) becomes

$$s_n(t) = \frac{2}{T} \int_0^t D_n\left(\frac{2\pi x}{T}\right) dx.$$

For convenience we now make the change of variable $u = 2\pi x/T$, which results in

$$s_n(t) = \frac{1}{\pi} \int_0^{2\pi t/T} D_n(u) \, du. \tag{5.54}$$

Equation (5.54) describes the behavior of the first n terms of the Fourier series for any and all values of t in the interval $[0, T/2]$. We note that when $t = T/2$, the upper limit on the integral becomes π. Thus, we see we need only consider the behavior of $D_n(u)$ (see Figure 5.2) up to the value $u = \pi$. Furthermore, because of the symmetric nature of $f(t)$ (an odd function) we can easily determine that the nth partial sum is given as $-s_n(t)$ for any t in the interval $[-T/2, 0]$. In Figure 5.4 we show the nth partial sum approximation to the function $f(t)$ for $n = 3$, 12, and 24, respectively. As can be appreciated

---

*Note here that $f(+)$ means f is evaluated at positive values of x; similarly, $f(-)$ indicates that we only have negative values of x.

from this figure, for large values of n the basic shape of the ringing remains the same, although it is compressed toward the point of discontinuity. As n approaches infinity, all the ringing is confined to a set of measure zero, that is, the point of discontinuity.

It is interesting to note that the frequency of the ringing, or Gibbs' effect, is $(2n + 1)\pi/T$, which can be directly correlated to the frequency $(2\pi n/T)$ of the last term(s) used in the Fourier series approximation. In other words, the frequency of the ringing is equal to that of the first term omitted from the Fourier series approximation.

For small values of t we can use the approximation

$$D_n(u) \sim \frac{\sin([n + 1/2]u)}{u}.$$

As it turns out, for $t < T/10$ the error in this approximation is less that 2% and for $t < T/20$ the error is less that 0.5%. When this approximation is substituted in equation (5.54), we obtain the following simplified formula:

$$s_n(t) \sim \frac{1}{\pi} \int_0^{2(n+1)\pi t/T} \frac{\sin(u)}{u} \, du \qquad \text{(small } t\text{)}. \qquad (5.55)$$

Shown in Figure 5.5 is a plot of equation (5.55) with $y = (2n + 1)\pi t/T$ ranging from 0 to 50. We see from this figure that $s_n(y)$ rises toward the limit value of $\frac{1}{2}$, overshoots it, and then oscillates about this limit value. We find (by differentiating under the integral) that the maxima and minima occur at $y = k\pi$ $(k = 1, 2, ...)$. From the graph we determine that at the first maximum $s_n(y) = 0.589$ and at the first minimum $s_n(y) = .452$. Thus, $s_n(y)$ overshoots the limiting value by approximately 18% and undershoots it by approximately 10%. If we consider the total magnitude of the jump from $-\frac{1}{2}$ to $+\frac{1}{2}$, these overshoot values become 9% and 5%, respectively. (The same remarks are also true for the function $D_n(y)$).

As our final topic in this section we consider the behavior of the Fourier series in the neighborhood of *any jump discontinuity*. For this situation we consider only a very small neighborhood of radius δ and a very large number of terms in the approximation. In this case we assume that the Localization Theorem (Theorem 5.12) is valid, and we can describe the nth partial sum as

$$s_n(t) = \frac{2}{T} \int_0^{\delta} [f(t + u) + f(t - u)] D_n\left(\frac{2\pi u}{T}\right) du.$$

Since (by assumption) δ is very small, we assume that on the left side of the jump $f(t - u) = f(t^-)$ can be considered a constant. Similarily, on the right side of the jump $f(t + u) = f(t^+)$. Consequently, the constant expression $f(t^+) + f(t^-)$ can be removed from the integral and we have

$$s_n(t) = \frac{2}{T} [f(t^+) + f(t^-)] \int_0^{\delta} D_n\left(\frac{2\pi u}{T}\right) du.$$

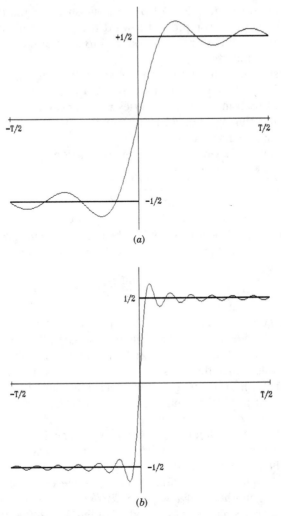

Figure 5.4 The nth partial sum approximation to jump discontinuity for (a) n= 3, (b) n= 12, and (c) n= 24.

Thus, we see that within this *arbitrarily small neighborhood* to the right of the discontinuity (for large n), the behavior of the Fourier series approximation is essentially the same as that presented for the function of equation (5.52). Similar remarks hold true for the behavior of the Fourier series in an arbitrarily small neighborhood to the left of the discontinuity. It is important to keep in mind that when we discussed Gibbs' effect for the function given by equation (5.52) we were able to describe it over the entire interval T. Here, for the general case, we must limit our attention to some arbitrarily small neighborhood of the jump discontinuity.

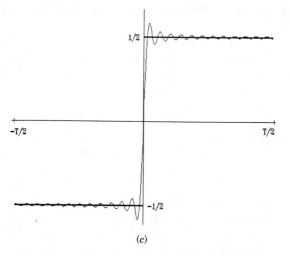

(c)

Figure 5.4   *(Continued)*

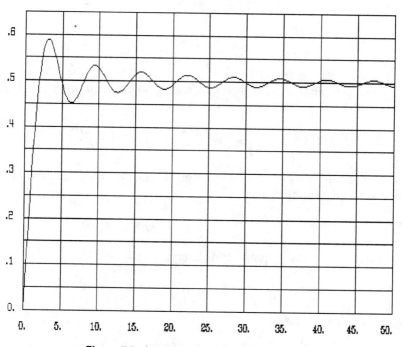

Figure 5.5   Integral of sinc(*y*)—equation (5.55).

FOURIER SERIES REPRESENTATION OF COMPLEX FUNCTIONS

Up to this point when dealing with the Fourier series of a function we have always implicitly assumed that the function was real. We now consider the situation when the function is complex with a real and an imaginary portion. Let us write a complex function $f(t)$ as

$$f(t) = f_R(t) + if_I(t),$$

where $f_R(t)$ is the real portion of the function and $f_I(t)$ is the imaginary portion. For example, consider the complex function

$$f(t) = e^{iat} = \cos(at) + i\sin(at).$$

In this case we obviously have $f_R(t) = \cos(at)$ and $f_I(t) = \sin(at)$.

To consider the Fourier series representation of a complex function $f(t)$ let us first make the assumption that both $f_R(t)$ and $f_I(t)$ are periodic over the interval T. In addition, let us also assume that both functions can be represented by Fourier series. In other words, we assume that the following equations are true:

$$f_R(t) = A_0 + \sum_{k=1}^{\infty} A_k \cos\left(\frac{2\pi kt}{T}\right) + B_k \sin\left(\frac{2\pi kt}{T}\right)$$

and

$$f_I(t) = C_0 + \sum_{k=1}^{\infty} C_k \cos\left(\frac{2\pi kt}{T}\right) + D_k \sin\left(\frac{2\pi kt}{T}\right).$$

Now let us assume that the complex function $f(t)$ can be written as a Fourier series, that is,

$$f(t) = \alpha_0 + \sum_{k=1}^{\infty} \alpha_k \cos\left(\frac{2\pi kt}{T}\right) + \beta_k \sin\left(\frac{2\pi kt}{T}\right).$$

Since (by assumption) the above equation is valid, we know that the Fourier series coefficients are given as

$$\alpha_0 = \frac{1}{T}\int_T f(t)\,dt = \frac{1}{T}\int_T (f_R(t) + if_I(t))\,dt.$$

However, by the linearity property of integrals we have

$$\alpha_0 = \frac{1}{T}\int_T f_R(t)\,dt + \frac{i}{T}\int_T f_I(t)\,dt = A_0 + iC_0.$$

Similar reasoning leads to

$$\alpha_k = A_k + iC_k, \qquad \beta_k = B_k + iD_k, \qquad k = 1, 2, \dots.$$

Thus, we see that the Fourier series coefficients of a complex function are simply linear combination of the Fourier series coefficients of the real and imaginary portions of the complex function.

When the complex function is expressed as a Fourier series using the complex exponential orthogonal system of equation (5.40), that is,

$$f(t) = \sum_{k=-\infty}^{\infty} C_k e^{2\pi ikt/T},$$

then it is also possible to demonstrate that

$$C_k = R_k + iS_k,$$

where R_k and S_k are the complex exponential Fourier series coefficients of the real and imaginary components of the function.

To determine pointwise convergence of the Fourier series of a complex function, we must examine the pointwise convergence of both its real and imaginary portions.

PROPERTIES OF THE FOURIER SERIES

In this section we discuss properties of the Fourier series that can often be used to simplify the calculation of the Fourier series coefficients of a function. We first consider linearity of the Fourier series representation of a function.

Theorem 5.13 (Linearity). If the functions $f(t)$ and $g(t)$ can both be represented by a Fourier series with coefficients $(A_0, A_k,$ and $B_k)$ and $(C_0, C_k,$ and $D_k)$, respectively, then the function $h(t) = af(t) + bg(t)$ can also be represented by a Fourier series whose coefficients are given as $aA_0 + bC_0$, $aA_k + bC_k$, and $aB_k + bD_k$.

Proof. By assumption,

$$h(t) = af(t) + bg(t),$$

$$h(t) = a\left[A_0 + \sum_{k=1}^{\infty} A_k \cos\left(\frac{2\pi kt}{T}\right) + B_k \sin\left(\frac{2\pi kt}{T}\right)\right]$$

$$+ b\left[C_0 + \sum_{k=1}^{\infty} C_k \cos\left(\frac{2\pi kt}{T}\right) + D_k \sin\left(\frac{2\pi kt}{T}\right)\right].$$

By regrouping terms, we obtain

$$h(t) = (aA_0 + bC_0) + \sum_{k=1}^{\infty}(aA_k + bC_k)\cos\left(\frac{2\pi kt}{T}\right)$$

$$+ \sum_{k=1}^{\infty}(aB_k + bD_k)\sin\left(\frac{2\pi kt}{T}\right).$$

Thus,

$$h(t) = \alpha_0 + \sum_{k=1}^{\infty} \alpha_k \cos\left(\frac{2\pi k t}{T}\right) + \beta_k \sin\left(\frac{2\pi k t}{T}\right).$$

We see therefore that $h(t)$ is represented by a Fourier series whose coefficients are given as per our supposition.   Q.E.D.

Corollary 5.13. If the functions $f(t)$ and $g(t)$ can both be represented over the interval T by a Fourier series of complex exponentials with coefficients C_k and D_k, respectively, then the function $h(t) = af(t) + bg(t)$ can also be represented by a Fourier series of complex exponentials whose coefficients are given as $aC_k + bD_k$.

Theorem 5.14 (First Shifting Theorem). If the function $f(t)$ can be represented by a Fourier series over the interval T with coefficients A_0, A_k, and B_k, then the function $f(t-a)$ can also be represented by a Fourier series with coefficients α_0, α_k, and β_k, where

$$\alpha_0 = A_0,$$

$$\alpha_k = A_k \cos\left(\frac{2\pi k a}{T}\right) - B_k \sin\left(\frac{2\pi k a}{T}\right),$$

$$\beta_k = A_k \sin\left(\frac{2\pi k a}{T}\right) + B_k \cos\left(\frac{2\pi k a}{T}\right).$$

Proof. By assumption,

$$f(t) = A_0 + \sum_{k=1}^{\infty} A_k \cos\left(\frac{2\pi k t}{T}\right) + B_k \sin\left(\frac{2\pi k t}{T}\right).$$

We now substitute $t - a$ for t in the above equation to obtain

$$f(t-a) = A_0 + \sum_{k=1}^{\infty} A_k \cos\left(\frac{2\pi k [t-a]}{T}\right) + B_k \sin\left(\frac{2\pi k [t-a]}{T}\right).$$

Now using equations (5.33) and (5.34) we rewrite the above equation as

$$f(t-a) = A_0 + \sum_{k=1}^{\infty} A_k \left[\cos\left(\frac{2\pi k t}{T}\right) \cos\left(\frac{2\pi k a}{T}\right) + \sin\left(\frac{2\pi k t}{T}\right) \sin\left(\frac{2\pi k a}{T}\right)\right]$$

$$+ \sum_{k=1}^{\infty} B_k \left[\sin\left(\frac{2\pi k t}{T}\right) \cos\left(\frac{2\pi k a}{T}\right) - \cos\left(\frac{2\pi k t}{T}\right) \sin\left(\frac{2\pi k a}{T}\right)\right].$$

By regrouping, we find

$$f(t - a) = A_0 + \sum_{k=1}^{\infty} \left[A_k \cos\left(\frac{2\pi k a}{T}\right) - B_k \sin\left(\frac{2\pi k a}{T}\right) \right] \cos\left(\frac{2\pi k t}{T}\right)$$

$$+ \sum_{k=1}^{\infty} \left[A_k \sin\left(\frac{2\pi k a}{T}\right) + B_k \cos\left(\frac{2\pi k a}{T}\right) \right] \sin\left(\frac{2\pi k t}{T}\right).$$

We now let

$$\alpha_0 = A_0,$$

$$\alpha_k = A_k \cos\left(\frac{2\pi k a}{T}\right) - B_k \sin\left(\frac{2\pi k a t}{T}\right),$$

and

$$\beta_k = A_k \sin\left(\frac{2\pi k a}{T}\right) + B_k \cos\left(\frac{2\pi k a}{T}\right),$$

which results in

$$f(t - a) = \alpha_0 + \sum_{k=1}^{\infty} \alpha_k \cos\left(\frac{2\pi k t}{T}\right) + \beta_k \sin\left(\frac{2\pi k t}{T}\right).$$

Thus, the function $f(t - a)$ is expressed as a Fourier series with coefficients α_0, α_k, and β_k, which are given as per our supposition.   Q.E.D.

Corollary 5.14. If the function $f(t)$ can be represented as a Fourier series over the interval T using complex exponentials with coefficients C_k, then the function $f(t - a)$ can also be represented as a Fourier series over the interval T with coefficients given as $C_k \exp[-2\pi i k a/T]$.

Theorem 5.15 (Second Shifting Theorem). If the function $f(t)$ can be represented over the interval T by a Fourier series with coefficients A_0, A_k, and B_k, then the function $h(t) = f(t)\exp[2\pi i a t/T]$ can also be represented by a Fourier series with coefficients α_0, α_k, and β_k, where

$$\alpha_0 = \frac{A_a}{2} + \frac{iB_a}{2},$$

$$\alpha_k = \frac{A_{k+a} + A_{k-a}}{2} + \frac{i(B_{k+a} - B_{k-a})}{2},$$

$$\beta_k = \frac{B_{k+a} + B_{k-a}}{2} - \frac{i(A_{k+a} - A_{k-a})}{2}.$$

Proof. Let us first briefly digress and note that by substituting $k + a$ and $k - a$ for k in equations (5.4) and (5.5) we obtain

$$A_{k+a} = \frac{2}{T} \int_T f(t) \cos \left(\frac{2\pi(k+a)t}{T} \right) dt,$$

$$A_{k-a} = \frac{2}{T} \int_T f(t) \cos \left(\frac{2\pi(k-a)t}{T} \right) dt,$$

$$B_{k+a} = \frac{2}{T} \int_T f(t) \sin \left(\frac{2\pi(k+a)t}{T} \right) dt,$$

$$B_{k-a} = \frac{2}{T} \int_T f(t) \sin \left(\frac{2\pi(k-a)t}{T} \right) dt.$$

Now, by assumption, we have

$$f(t) = A_0 + \sum_{k=1}^{\infty} A_k \cos \left(\frac{2\pi k t}{T} \right) + B_k \sin \left(\frac{2\pi k t}{T} \right).$$

Next we multiply both sides of the above equation by

$$\exp[2\pi at/T] = \cos(2\pi at/T) + i \sin(2\pi at/T)$$

to obtain

$$f(t)e^{2\pi iat/T} = A_0 \left[\cos \left(\frac{2\pi at}{T} \right) + i \sin \left(\frac{2\pi at}{T} \right) \right]$$

$$+ \sum_{k=1}^{\infty} A_k \left[\cos \left(\frac{2\pi at}{T} \right) + i \sin \left(\frac{2\pi at}{T} \right) \right] \cos \left(\frac{2\pi k t}{T} \right)$$

$$+ \sum_{k=1}^{\infty} B_k \left[\cos \left(\frac{2\pi at}{T} \right) + i \sin \left(\frac{2\pi at}{T} \right) \right] \sin \left(\frac{2\pi k t}{T} \right)$$

or

$$f(t)e^{2\pi iat/T} = \alpha_0 + \sum_{k=1}^{\infty} \alpha_k \cos \left(\frac{2\pi k t}{T} \right) + \beta_k \sin \left(\frac{2\pi k t}{T} \right),$$

where

$$\alpha_0 = A_0 \left(\cos \left(\frac{2\pi at}{T} \right) + i \sin \left(\frac{2\pi at}{T} \right) \right),$$

$$\alpha_k = A_k \left(\cos \left(\frac{2\pi at}{T} \right) + i \sin \left(\frac{2\pi at}{T} \right) \right),$$

$$\beta_k = B_k \left(\cos \left(\frac{2\pi at}{T} \right) + i \sin \left(\frac{2\pi at}{T} \right) \right).$$

Let's first consider the above expression for α_k, that is,

$$\alpha_k = A_k \left(\cos \left(\frac{2\pi at}{T} \right) + i \sin \left(\frac{2\pi at}{T} \right) \right),$$

or, using equation (5.4), we substitute the integral expression for A_k to obtain

$$\alpha_k = \frac{2}{T} \int_T f(t) \cos \left(\frac{2\pi kt}{T} \right) \cos \left(\frac{2\pi at}{T} \right) dt$$

$$+ \frac{2i}{T} \int_T f(t) \cos \left(\frac{2\pi kt}{T} \right) \sin \left(\frac{2\pi at}{T} \right) dt.$$

Now using equations (5.35) and (5.37) we find

$$\alpha_k = \frac{1}{T} \int_T f(t) \cos \left(\frac{2\pi(k+a)t}{T} \right) dt + \frac{1}{T} \int_T f(t) \cos \left(\frac{2\pi(k-a)t}{T} \right) dt$$

$$+ \frac{i}{T} \int_T f(t) \sin \left(\frac{2\pi(k+a)t}{T} \right) dt - \frac{i}{T} \int_T f(t) \sin \left(\frac{2\pi(k-a)t}{T} \right) dt,$$

or

$$\alpha_k = \frac{A_{k+a} + A_{k-a}}{2} + \frac{i[B_{k+a} - B_{k-a}]}{2}.$$

Similar reasoning yields the desired results for α_0 and β_k.   Q.E.D.

Corollary 5.15a. If the real function $f(t)$ can be represented over the interval T by a Fourier series with coefficients A_0, A_k, and B_k, then:

(i) The function $f(t)\cos(2\pi at/T)$ can also be represented by a Fourier series over the interval T whose coefficients are A_a, $[A_{k+a} + A_{k-a}]/2$, and $[B_{k+a} + B_{k-a}]/2$.

(ii) The function $f(t)\sin(2\pi at/T)$ can also be represented by a Fourier series over the interval T whose coefficients are B_a, $[B_{k+a} - B_{k-a}]/2$, and $[A_{k+a} - A_{k-a}]/2$.

Proof. The proof of this corollary follows directly from Theorem 5.15 by equating the real and imaginary portions of the function with the real and imaginary portions of the series representation.   Q.E.D.

Corollary 5.15b. If the function $f(t)$ can be represented as a Fourier series over the interval T using the complex exponential system with coefficients C_k, then the function $f(t)\exp[2\pi iat/T]$ can also be represented by a complex exponential Fourier series with coefficients C_{k-a}.

We now consider the situation where the function $f(t)$ is either odd or even. A function is called *even* if and only if $f(-t) = f(t)$; similarly, it is called *odd* if and only if $f(-t) = -f(t)$. Note that by their very nature, odd and even functions are centered (about 0) in the middle of the interval T.

Theorem 5.16 (Even Function). If $f(t)$ is an even function that can be represented over the interval $T = [-T/2, T/2]$ by a Fourier series, then the coefficients are given as

$$A_0 = \frac{2}{T} \int_0^{T/2} f(t)\,dt,$$

$$A_k = \frac{4}{T} \int_0^{T/2} f(t)\cos\left(\frac{2\pi k t}{T}\right) dt, \qquad k = 1,\ldots,$$

$$B_k = 0, \qquad \text{for all } k.$$

Proof. By definition,

$$A_0 = \frac{1}{T} \int_{-T/2}^{T/2} f(t)\,dt.$$

We now split the integral and write

$$A_0 = \frac{1}{T} \int_{-T/2}^{0} f(t)\,dt + \frac{1}{T} \int_0^{T/2} f(t)\,dt.$$

If we now substitute $-t$ for t (and $-dt$ for dt) in the first integral, we obtain

$$A_0 = -\frac{1}{T} \int_{T/2}^{0} f(-t)\,dt + \frac{1}{T} \int_0^{T/2} f(t)\,dt.$$

Now inverting the limits on the first integral (which changes the sign of the integral) and noting that for an even function $f(-t) = f(t)$, we find

$$A_0 = \frac{1}{T} \int_0^{T/2} f(t)\,dt + \frac{1}{T} \int_0^{T/2} f(t)\,dt,$$

or

$$A_0 = \frac{2}{T} \int_0^{T/2} f(t)\,dt.$$

Noting that the product $f(t)\cos(2\pi k t/T)$ is also an even function, we can use the same type of logic to obtain the result for the A_k coefficients.

Now for the B_k coefficients, we have

$$B_k = \frac{2}{T} \int_{-T/2}^{T/2} f(t) \sin \left(\frac{2\pi k t}{T} \right) dt,$$

or

$$B_k = \frac{2}{T} \int_{-T/2}^{0} f(t) \sin \left(\frac{2\pi k t}{T} \right) dt + \frac{2}{T} \int_{0}^{T/2} f(t) \sin \left(\frac{2\pi k t}{T} \right) dt.$$

Again, substitution of $-t$ for t and $-dt$ for dt in the first integral of the previous equation yields

$$B_k = -\frac{2}{T} \int_{T/2}^{0} f(-t) \sin \left(-\frac{2\pi k t}{T} \right) dt + \frac{2}{T} \int_{0}^{T/2} f(t) \sin \left(\frac{2\pi k t}{T} \right) dt.$$

We invert the limits on the first integral and recall that for the even function we have $f(-t) = f(t)$ and that for the odd function we have $\sin(-2\pi k t/T) = -\sin(2\pi k t/T)$. This results in

$$B_k = -\frac{2}{T} \int_{0}^{T/2} f(t) \sin \left(\frac{2\pi k t}{T} \right) dt + \frac{2}{T} \int_{0}^{T/2} f(t) \sin \left(\frac{2\pi k t}{T} \right) dt,$$

or

$$B_k = 0. \qquad\qquad \text{Q.E.D.}$$

The evaluation of the Fourier series coefficients is significantly simplified for an even function. All the B_k coefficients vanish, and the A_k terms are now calculated by performing the integration from 0 to $T/2$ (rather than from $-T/2$ to $T/2$) and then doubling the value of the results.

Using analogous reasoning we can prove the following theorem.

Theorem 5.17 (Odd Function). If $f(t)$ is an odd function that can be represented over the interval $T = [-T/2, T/2]$ by a Fourier series, then the coefficients are given as

$$B_k = \frac{4}{T} \int_{0}^{T/2} f(t) \sin \left(\frac{2\pi k t}{T} \right) dt, \qquad k = 1, \ldots,$$

$$A_k = 0, \qquad \text{for all } k.$$

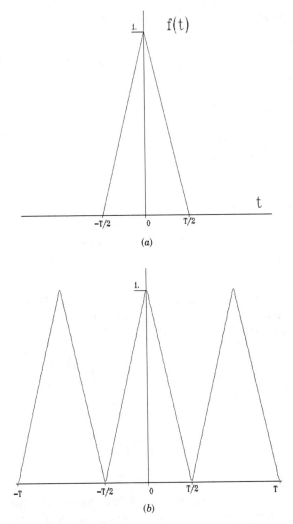

Figure 5.6 (a) Triangle function and (b) its 20-term Fourier series representation.

EXAMPLES

In this section we present several example problems that should help illustrate the material and concepts presented in this chapter.

Example 1. We begin with the triangle function that is shown in Figure 5.6a and described mathematically as

$$f(t) = \begin{cases} 1 + 2t/T, & t \in [-T/2, 0), \\ 1, & t = 0, \\ 1 - 2t/T, & t \in (0, T/2]. \end{cases} \tag{5.56}$$

We first note that this function is even; consequently, $B_k = 0$ for all k. Furthermore, we have

$$A_0 = \frac{2}{T} \int_0^{T/2} f(t)\,dt = \frac{2}{T} \int_0^{T/2} \left(1 - \frac{2t}{T}\right) dt = \frac{2}{T} \left(t - \frac{t^2}{T}\right) = \frac{1}{2}.$$

$$A_k = \frac{4}{T} \int_0^{T/2} f(t) \cos\left(\frac{2\pi k t}{T}\right) dt = \frac{4}{T} \int_0^{T/2} \left(1 - \frac{2t}{T}\right) \cos\left(\frac{2\pi k t}{T}\right) dt,$$

$$= \frac{4}{T} \int_0^{T/2} \cos\left(\frac{2\pi k t}{T}\right) dt - \frac{8}{T^2} \int_0^{T/2} t \cos\left(\frac{2\pi k t}{T}\right) dt,$$

$$A_k = \frac{4}{T} \frac{\sin(2\pi k t/T)}{2\pi k/T}\Big|_0^{T/2} - \frac{8}{T^2} \left[\frac{\cos(2\pi k t/T)}{(2\pi k/T)^2} + \frac{\sin(2\pi k t/T)}{2\pi k/T}\right]\Big|_0^{T/2},$$

$$A_k = \frac{2 - 2\cos(\pi k)}{(\pi k)^2}.$$

Thus,

$$f(t) = \frac{1}{2} + \sum_{k=1}^{\infty} \frac{2 - 2\cos(\pi k)}{(\pi k)^2} \cos\left(\frac{2\pi k t}{T}\right).$$

The first 20 terms of the Fourier series representation of this function are shown in Figure 5.6b.

Example 2. In this example we consider the function defined mathematically as

$$f(t) = \begin{cases} 1 + 2t/T, & t \in [-T/2, 0), \\ 0, & t = 0, \\ 1 - 2t/T, & t \in (0, T/2]. \end{cases} \tag{5.57}$$

This function is very similar to the one defined in equation (5.56). The only difference is the behavior of the function at $t = 0$. That is to say, equation (5.57) describes a triangle function with a hole at the origin (see Figure 5.7a).

Again, we see that this is an even function and we use Theorem 5.16 to simplify the evaluation of the Fourier series coefficients, that is,

$$A_0 = \frac{2}{T} \int_0^{T/2} \left(1 - \frac{2t}{T}\right) dt$$

and

$$A_k = \frac{4}{T} \int_0^{T/2} \left(1 - \frac{2t}{T}\right) \cos\left(\frac{2\pi k t}{T}\right) dt.$$

We note, however, that the above integrals are taken in the Lebesgue sense and are not affected by the behavior of the function on a set of measure zero, for example, $t = 0$. Consequently, we obtain the same coefficients as in the previous example. In other words, functions (5.56) and (5.57) have the same

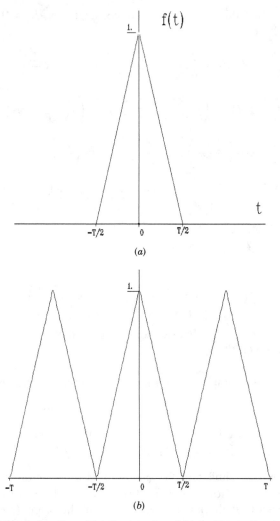

Figure 5.7   (a) Triangle function with hole at $t = 0$ and (b) its 20-term Fourier series approximation.

Fourier series representation. This should not be too surprising inasmuch as the two functions are equal *almost everywhere* over the interval T, differing only at the point $t = 0$.

Let us now examine the pointwise convergence of the Fourier series at $t = 0$. First, for the function defined by equation (5.56) (Figure 5.6a) we note

$$f(0^+) = \lim_{h \to 0} f(0 + h) = 1$$

and

$$f(0^-) = \lim_{h \to 0} f(0 - h) = 1.$$

Thus, equation (5.51) tells us that the series converges to 1 at $t = 1$.

Now let's look at the local behavior of the function described by equation (5.57) (see Figure 5.7a). Even though $f(0) = 0$ we note that

$$f(0^+) = \lim_{h \to 0} f(0 + h) = 1$$

and

$$f(0^-) = \lim_{h \to 0} f(0 - h) = 1.$$

Again we use equation (5.51) to show that the Fourier series representation of this function converges to 1 at $t = 0$, even though $f(0) = 0$.

Example 3. Consider the function described mathematically as

$$f(t) = \begin{cases} -2t/T, & t \in [-T/2, 0), \\ 2t/T, & t \in [0, T/2]. \end{cases} \tag{5.58}$$

This function is shown in Figure 5.8a. Again we are dealing with an even function; thus,

$$A_0 = \frac{2}{T} \int_0^{T/2} \frac{2t}{T} \, dt = \frac{2}{T^2} t^2 \Big|_0^{T/2} = \frac{1}{2}.$$

$$A_k = \frac{4}{T} \int_0^{T/2} \frac{2t}{T} \cos\left(\frac{2\pi k t}{T}\right) dt = \frac{2\cos(\pi k) - 2}{(\pi k)^2}.$$

Thus,

$$f(t) = \frac{1}{2} + \sum_{k=1}^{\infty} \frac{2\cos(\pi k) - 2}{(\pi k)^2} \cos\left(\frac{2\pi k t}{T}\right).$$

The first 20 terms of this Fourier series were used to generate the approximation shown in Figure 5.8b.

When comparing Figures 5.8 and 5.7 we see that in reality the function given by equation (5.58) can be considered the same as the function of equation (5.56) shifted to the right by an amount $a = T/2$. Consequently, we could have used the First Shifting Theorem (Theorem 5.14) to obtain the coefficients directly. In other words, let α_0, α_k, and β_k be the Fourier series coefficients of the function (5.58) and let A_0, A_k, and B_k be those of function (5.56). Then,

$$\alpha_0 = A_0 = \tfrac{1}{2},$$

$$\alpha_k = A_k \cos\left(\frac{2\pi k a}{T}\right) - B_k \sin\left(\frac{2\pi k a}{T}\right) = \left(\frac{2 - 2\cos(\pi k)}{(\pi k)^2}\right) \cos(\pi k),$$

$$\beta_k = A_k \sin\left(\frac{2\pi k a}{T}\right) + B_k \cos\left(\frac{2\pi k a}{T}\right) = 0.$$

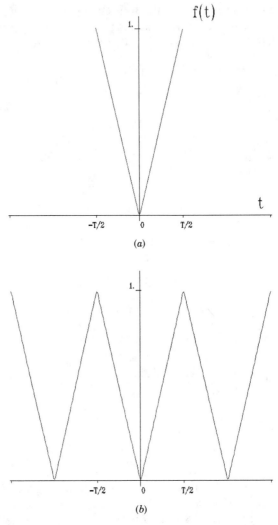

Figure 5.8 (a) "*V*" function and (b) its 20-term Fourier series approximation.

However, we note

$$[2 - 2\cos(\pi k)]\cos(\pi k) = 2\cos(\pi k) - 2\cos^2(\pi k)$$
$$= 2\cos(\pi k) - 2(1 - \sin^2(\pi k)),$$

or

$$[2 - 2\cos(\pi k)]\cos(\pi k) = 2\cos(\pi k) - 2.$$

Therefore,

$$\alpha_k = \frac{2\cos(\pi k) - 2}{(\pi k)^2},$$

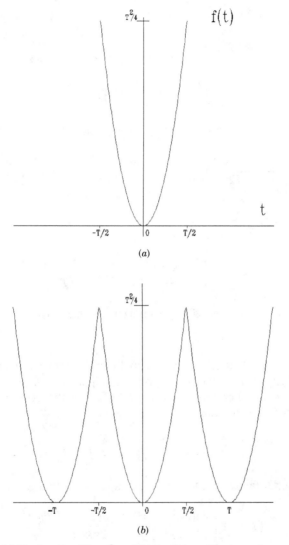

Figure 5.9 (a) The function $f(t) = t^2$ and (b) its 20-term Fourier series approximation.

which is exactly the same result that we obtained using the integral definition of the coefficients.

Example 4. Next we consider the function $f(t) = t^2$ over the interval $T = [-T/2, T/2]$. This function is shown in Figure 5.9a. Again we have an even function and thus, its Fourier series coefficients are given as:

$$A_0 = \frac{2}{T} \int_0^{T/2} f(t)\,dt = \frac{2}{T} \int_0^{T/2} t^2\,dt = \frac{T^2}{12}.$$

$$A_k = \frac{4}{T} \int_0^{T/2} t^2 \cos\left(\frac{2\pi k t}{T}\right) dt.$$

From an integral table, or using integration by parts twice, we obtain

$$A_k = \frac{4}{T} \left[\frac{2t\cos(2\pi k t/T)}{(2\pi k/T)^2} + \frac{(2\pi k/T)^2 t^2 - 2}{(2\pi k/T)^3} \sin\left(\frac{2\pi k t}{T}\right) \right]\Bigg|_0^{T/2}$$

$$A_k = \frac{\cos(\pi k)}{(\pi k/T)^2}.$$

Thus,

$$f(t) = \frac{T^2}{12} + \sum_{k=1}^{\infty} \frac{\cos(\pi k)}{(\pi k/T)^2} \cos\left(\frac{2\pi k t}{T}\right).$$

Shown in Figure 5.9b is the 20-term Fourier series approximation to the function $f(t) = t^2$.

Example 5. We next consider the ramp function $f(t) = t$ over the interval $T = [-T/2, T/2]$. This function is shown in Figure 5.10a. We note that this function is odd; and therefore, Theorem 5.17 tells us that A_0 and A_k are equal to zero for all k. Furthermore,

$$B_k = \frac{4}{T} \int_0^{T/2} f(t) \sin\left(\frac{2\pi k t}{T}\right) dt = \frac{4}{T} \int_0^{T/2} t \sin\left(\frac{2\pi k t}{T}\right) dt.$$

Using an integral table, or integration by parts, we obtain

$$B_k = \frac{4}{T} \left[\frac{\sin(2\pi k t/T)}{(2\pi k/T)^2} - \frac{t\cos(2\pi k t/T)}{2\pi k/T} \right]\Bigg|_0^{T/2}$$

$$B_k = -\frac{\cos(\pi k)}{\pi k}.$$

Thus,

$$f(t) = \sum_{k=1}^{\infty} -\frac{\cos(\pi k)}{\pi k} \sin\left(\frac{2\pi k t}{T}\right).$$

The 20-term Fourier series approximation to this function is shown in Figure 5.10b. Note the ringing, or Gibbs' effect, at the endpoints which is caused by the jump discontinuities. Also note that at this jump the Fourier series converges to the average values of the function just before the jump $(-T/2)$ and

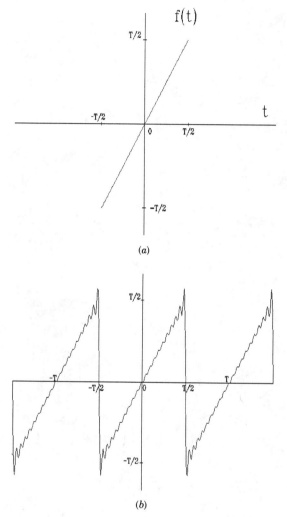

Figure 5.10   (a) The ramp function $f(t) = t$ and (b) its 20-term Fourier series representation.

just after the jump $(T/2)$. That is to say, at the jump the Fourier series converges to a value of zero. Furthermore, the frequency of the Gibbs' ringing (w_G) is equal to that of the first term omitted in the Fourier series approximation (in this case the 21st term). Thus $w_G = 21/T$.

Example 6. Consider the function shown in Figure 5.11a and defined mathematically over the interval $T = [-T/2, T/2]$ as

$$f(t) = \begin{cases} \cos(2\pi t/T), & t \in [-T/4, T/4], \\ 0, & \text{otherwise.} \end{cases}$$

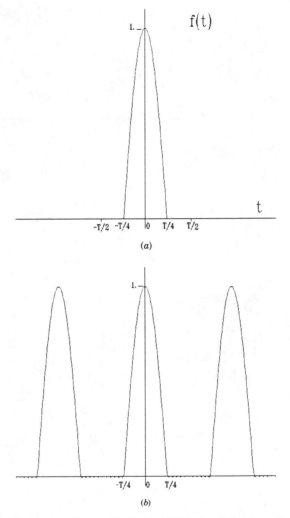

Figure 5.11 (a) The function $f(t) = \cos(2\pi t/T)$ and (b) its 20-term Fourier series approximation.

Once again we are dealing with an even function, and its coefficients are determined as follows:

$$A_0 = \frac{2}{T} \int_0^{T/2} f(t)\,dt = \frac{2}{T} \int_0^{T/4} \cos\left(\frac{2\pi t}{T}\right)dt,$$

$$A_0 = \frac{2}{T} \left. \frac{\sin(2\pi t/T)}{2\pi/T} \right|_0^{T/4} = \frac{1}{\pi},$$

$$A_k = \frac{4}{T} \int_0^{T/2} f(t)\cos\left(\frac{2\pi k t}{T}\right)dt = \frac{4}{T} \int_0^{T/4} \cos\left(\frac{2\pi t}{T}\right)\cos\left(\frac{2\pi k t}{T}\right)dt.$$

Using equation (5.35) we can express the above integral as

$$A_k = \frac{2}{T} \int_0^{T/4} \cos\left(\frac{2\pi(k+1)t}{T}\right) dt + \frac{2}{T} \int_0^{T/4} \cos\left(\frac{2\pi(k-1)t}{T}\right) dt.$$

We first note a special situation when $k = 1$ inasmuch as the cosine term in the second integral is a constant equal to 1. Thus,

$$A_1 = \left.\frac{\sin(4\pi t/T)}{2\pi}\right|_0^{T/4} + \left(\frac{2}{T}\right)\left(\frac{T}{4}\right) = 0 + \frac{1}{2} = \frac{1}{2}.$$

Now for all $k > 1$ we have

$$A_k = \left(\frac{2}{T}\right)\left.\frac{\sin(2\pi(k+1)t/T)}{2\pi(k+1)/T}\right|_0^{T/4} + \left(\frac{2}{T}\right)\left.\frac{\sin(2\pi(k-1)t/T)}{2\pi(k-1)/T}\right|_0^{T/4}$$

$$A_k = \frac{\sin(\pi(k+1)/2)}{\pi(k+1)} + \frac{\sin(\pi(k-1)/2)}{\pi(k-1)}.$$

Thus,

$$f(t) = \frac{1}{\pi} + \frac{1}{2}\cos\left(\frac{2\pi t}{T}\right)$$

$$+ \sum_{k=2}^{\infty}\left[\frac{\sin(\pi(k+1)/2)}{\pi(k+1)} + \frac{\sin(\pi(k-1)/2)}{\pi(k-1)}\right]\cos\left(\frac{2\pi k t}{T}\right).$$

The first 20 terms of this Fourier series approximation are shown in Figure 5.11b.

Example 7. As our final example, consider the function $f(t) = \exp[-at]$ over the interval $T = [0,T]$. This function is shown in Figure 5.12a. The Fourier series coefficients for this function are determined as follows:

$$A_0 = \frac{1}{T}\int_0^T f(t)dt = \frac{1}{T}\int_0^T e^{-at}\,dt = \left(-\frac{1}{aT}\right)[e^{-aT} - e^0],$$

$$A_0 = \frac{1 - e^{-aT}}{aT},$$

$$A_k = \frac{2}{T}\int_0^T f(t)\cos\left(\frac{2\pi k t}{T}\right)dt = \frac{2}{T}\int_0^T e^{-at}\cos\left(\frac{2\pi k t}{T}\right)dt.$$

Using an integral table or integration by parts, we find

$$A_k = \left(\frac{2}{T}\right)e^{-at}\left.\frac{-a\cos(2\pi k t/T) + (2\pi k/T)\sin(2\pi k t/T)}{a^2 + (2\pi k/T)^2}\right|_0^T$$

$$A_k = \left(\frac{2a}{T}\right)\frac{1 - e^{-aT}}{a^2 + (2\pi k/T)^2}.$$

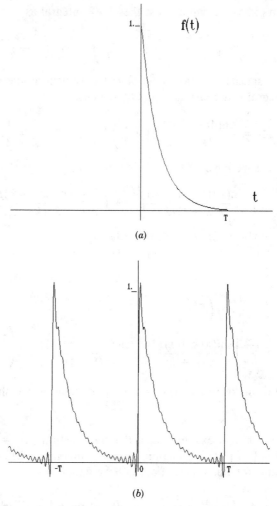

Figure 5.12 (a) The function $f(t) = e^{-at}$ and (b) and its 20-term Fourier series approximation.

Using similar reasoning, we determine the Fourier series sine coefficients to be given as

$$B_k = \left(\frac{4\pi k}{T^2}\right) \frac{1 - e^{-aT}}{a^2 + (2\pi k /T)^2}.$$

The Fourier series approximation to this function using 20 terms is shown in Figure 5.12b. Again, note the ringing, or Gibbs' effect, caused by the discontinuities at the edges of the interval.

SUMMARY

In this chapter we discussed the Fourier series representation of a periodic function. Although our main goal was to consider the trigonometric Fourier series, we first presented a more general discussion of Fourier series in terms of any orthogonal system of functions. Using this general approach we were able to consider such topics as convergence in the mean, complete systems, and the best least squares fit approximation of a function by a *finite* Fourier series. After this general presentation we limited our attention to the Fourier series representation of a function in terms of the trigonometric and complex exponential system of orthogonal functions, with particular emphasis placed on the trigonometric system. We presented the Dirichlet kernel function, which is used to study pointwise convergence of the Fourier series as well as Gibbs' effect or the behavior of a finite Fourier series at, and near, a point of discontinuity. Next a discussion of the properties of the Fourier series was presented. This presentation was particularly useful inasmuch as it provided the reader with techniques that can often be used to simplify the integral calculation of the Fourier series coefficients. Finally we ended this chapter with several example calculations to illustrate the concepts and techniques presented earlier.

PROBLEMS

1 Demonstrate that the sum, difference, product, and quotient of two periodic functions are also periodic.

2 Add and subtract $\sum c_k^2 \|O_k\|^2$ to both sides of equation (5.22) and perform the necessary algebraic manipulations required to obtain equation (5.23).

3 Prove the inequality

$$(f + g)^2 \le 2(f^2 + g^2).$$

(*Hint*: Add together the expanded expressions for $(f + g)^2$ and $(f - g)^2$.

4 Using equation (5.36) show that for $n \ne m$ we have

$$\int_T \sin\left(\frac{2\pi mt}{T}\right) \sin\left(\frac{2\pi nt}{T}\right) dt = 0.$$

5 Use the derivation of equation (5.13) as a guide and supply the proof of Theorem 5.6.

6 Supply the proof of Theorem 5.7 (use the derivation of equation (5.24) as a guide).

7 Supply the proof of Theorem 5.8 (use as a guide the derivation of equation 5.25 along with the fact that $\|O_0\|^2 = T$ and $\|O_k\|^2 = T/2$).

8 Multiply both sides of equation (5.6) by $\exp[-2\pi int/T]$ and apply the orthogonality conditions of equations (5.41) and (5.42) to obtain equation (5.7).

9 For the trigonometric system of equation (5.32) demonstrate the property of uniqueness (use the proof of Theorem 5.2 as a guide).

10 Use the derivation of equation (5.24) as a guide to show that if a square-summable function is approximated by the finite series

$$\sum_{k=-n}^{n} C_k e^{-2\pi ikt/T},$$

then the best approximation to the function (in terms of the mean square error) is obtained when the coefficients are calculated as per equation (5.7).

11 Establish Parseval's inequality for the complex exponential system of equation (5.40).

12 Using differentiation, show that the Dirichlet kernel function takes on its maximum values at $x = 2\pi k$.

13 Using L'Hôpital's rule show that $D_n(0) = (2n + 1)/2$.

14 Consider the function

$$s_n(y) = \int_0^y \frac{\sin(u)}{u}\,du.$$

Show that the maximums and minimums occur at $y = k\pi$, $(k = 1, 2, ...)$. Also determine the value of $s_n(y)$ at its first maximum and its first minimum.

15 Using the proof of Theorem 5.13 as a guide, supply the proof to Corollary 5.13.

16 Using the proof of Theorem 5.14 as a guide, supply the proof to Corollary 5.14.

17 Using the proof of Theorem 5.15 as a guide, supply the proofs of Corollaries 5.15a and 5.15b.

18 Supply the proof of Theorem 5.17 (use the proof of Theorem 5.16 as a guide).

19 Determine the complex exponential Fourier series representation of the triangle function of equation (5.56).

20 Determine the complex exponential Fourier series representation of the "V" function of equation (5.58).

21 Determine the complex exponential Fourier series representation of the function $f(t) = t^2$ over the interval $T = [-T/2, T/2]$.

22 Determine the complex exponential Fourier series representation of the function $f(t) = t$ over the interval $T = [-T/2, T/2]$.

23 Determine the complex exponential Fourier series representation of the function $f(t) = \cos(2\pi t/T)$ over the same domain used in Example 6.

24 Determine the complex exponential Fourier series representation of the function $f(t) = \exp[-at]$ over the interval $T = [0, T]$.

25 Determine the complex exponential Fourier series representation of the *boxcar* function defined over the interval $T = [-T/2, T/2]$ as

$$f(t) = \begin{cases} 1, & t \in [-T/4, T/4], \\ 0, & \text{otherwise.} \end{cases}$$

26 Determine the (trigonometric) Fourier series representation of the function $f(t) = t$ over the interval $T = [1, 2]$.

27 Determine the Fourier series representation of the function

$$f(t) = \begin{cases} \sin(2\pi t) & 0 \le t \le \frac{1}{2}, \\ 0, & \text{otherwise,} \end{cases}$$

over the interval $T = [0, 1]$.

28 Determine the Fourier series representation of the function $\sin(2\pi a t)$ over the interval $T = [0, 1]$ for the situation where a is not an integer.

29 Expand the function $f(t) = t^2$ in a Fourier series over the interval $T = [0, 1]$.

30 Expand the function $f(t) = at^2 + bt$ in a Fourier series over the interval $T = [-1, 1]$.

31 Expand the function $f(t) = |t|$ in a Fourier series over the interval $T = [-1, 1]$.

BIBLIOGRAPHY

Churchill, R. V., *Fourier Series and Boundary Value Problems*, McGraw-Hill, New York, 1963.

Hamming, R. W., *Digital Filters*, Prentice-Hall, Englewood Cliffs, N.J., 1977.

Ritt, R. V., *Fourier Series*, McGraw-Hill, New York, 1970.

Tolstov, G. P., *Fourier Series*, Prentice-Hall, Englewood Cliffs, N.J., 1962.

Weaver, H. J., *Applications of Discrete and Continuous Fourier Analysis*, John Wiley & Sons, New York, 1983.

6

THE FOURIER TRANSFORM

In the previous chapter we discussed the Fourier series, or, in other words, the frequency domain representation of a periodic function. In this chapter we consider the frequency domain representation of functions that are not necessarily required to be periodic. To accomplish this we use the Fourier transform.

Given a function $f(t)$ defined over the real line $R = (-\infty, \infty)$, we define the *Fourier transfrom pair* as

$$F(w) = \int_R f(t)e^{-2\pi i w t}\, dt, \qquad (6.1)$$

$$f(t) = \int_R F(w)e^{2\pi i w t}\, dw. \qquad (6.2)$$

Equation (6.1) is called the (direct) Fourier transform, and equation (6.2) is known as the inverse Fourier transform. We say $F(w)$ is the Fourier transform of $f(t)$ and that $f(t)$ is the inverse Fourier transform of $F(w)$. Notationally we write

$$F(w) = \mathcal{F}[f(t)] \quad \text{and} \quad f(t) = \mathcal{F}^{-1}[F(w)].$$

We point our here that the integrals in equations (6.1) and (6.2) are to be considered in the Lebesgue sense. Furthermore, the function $f(t)$ may be complex (although most our work in this chapter will be with real functions). We also note that the interval R refers to the entire real line and

$$\int_R g(t)\, dt = \int_{-\infty}^{\infty} g(t)\, dt.$$

EXISTENCE OF THE FOURIER TRANSFORM

In this section we consider the question of existence of the Fourier transform of a function $f(t)$. In other words, we wish to determine the conditions that must be placed on $f(t)$ in order to guarantee the existence of the integral of equation (6.1).

Before we proceed, let us briefly digress and recall from Chapter 3 that if the function f is Lebesgue integrable over the finite interval $[a,b]$, and if there exists a positive constant M such that

$$\int_a^b |f(t)|\,dt < M \qquad \text{for all} \quad b > a, \tag{6.3}$$

then f is Lebesgue integrable over the entire real line (i.e., $f \in L(R)$) and, in fact,

$$\int_R f(t)\,dt = \int_{-\infty}^{\infty} f(t)\,dt = \lim_{a \to \infty} \int_a^c f(t)\,dt + \lim_{b \to \infty} \int_c^b f(t)\,dt. \tag{6.4}$$

We note that functions that satisfy equation (6.3) are called *absolutely integrable* over the interval $[a,b]$ and are denoted as $|f| \in L([a,b])$. Therefore, when we say $|f| \in L(R)$, we mean that the function is such that equation (6.3) is valid. That is to say, $|f| \in L(R)$ implies that there exists a positive constant, M such that

$$\int_R |f(t)|\,dt < M.$$

It is worthwhile noting that in equation (6.3) the upper and lower limits are allowed to approach infinity independently of each other. When we require that the limits approach infinity while remaining equal we refer to this as the *principal value* (PV) of the integral. Also note that, by definition, $|f| \in L(R)$ implies that $f \in L(R)$.

Based on the above comments we are able to prove the following theorem.

Theorem 6.1 (Existence). If the function $f(t)$ is absolutely integrable over the real line, then its Fourier transform exists and, in addition, this transform is uniformly bounded.

Proof. We first note that

$$|f(t)e^{-2\pi i wt}| \le |f(t)||e^{-2\pi i wt}| = |f(t)|,$$

and since $|f| \in L(R)$ we have

$$\int_R |f(t)e^{-2\pi i wt}|\,dt \le \int_R |f(t)|\,dt < M,$$

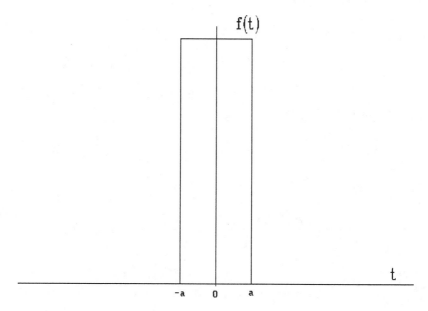

Figure 6.1 The unit pulse function.

which implies that the above integral, and consequently the Fourier transform $F(w)$, exists. The fact that $F(w)$ is uniformly bounded follows directly from Theorem 3.4(d), that is,

$$|F(w)| = \left| \int_R f(t) e^{-2\pi i w t} \, dt \right| \leq \int_R |f(t) e^{-2\pi i w t}| \, dt < M. \qquad \text{Q.E.D.}$$

Example 1. As our first example in this chapter, let us determine the Fourier transform of the unit pulse function of half-width a, which is defined as

$$p_a(t) = \begin{cases} 1, & |t| \leq a, \\ 0, & \text{otherwise.} \end{cases} \qquad (6.5)$$

This function is shown in Figure 6.1. We first note that

$$\int_R |p_a(t)| \, dt = 2a,$$

and consequently $|p_a(t)| \in L(R)$. Thus, it makes sense to calculate its Fourier transform using equation (6.1) as we did in Chapter 1 to obtain

$$\mathcal{F}[p_a(t)] = 2a \operatorname{sinc}(2\pi a w), \qquad (6.6)$$

which is shown in Figure 6.2.

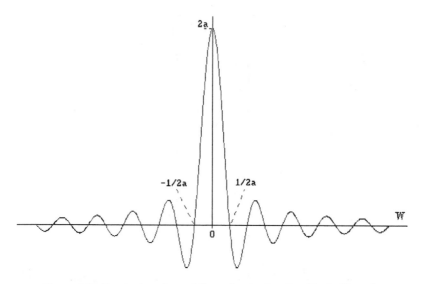

Figure 6.2 Fourier transform of the unit pulse function ($2a\,\text{sinc}[2\pi aw]$).

Example 2. We next calculate the Fourier transform of the one-sided exponential function, which is defined as

$$f(t) = \begin{cases} e^{-at}, & t \geq 0, \\ 0, & \text{otherwise.} \end{cases} \tag{6.7}$$

This function is shown in Figure 6.3. Note: a is a real constant greater than zero.

We again note that

$$\int_R |f(t)|\,dt = \int_R e^{-at}\,dt = \frac{1}{a},$$

and thus $|e^{-at}| \in L(R)$. Therefore it makes sense to calculate its Fourier tranform as per equation (6.1). When this is accomplished (see Chapter 1) we obtain

$$F(w) = \frac{1}{a + 2\pi i w} = \frac{a - 2\pi i w}{a^2 + 4\pi^2 w^2}. \tag{6.8}$$

In this situation, we note that even though the exponential function of equation (6.7) is real, its Fourier transform is complex. Shown in Figure 6.4 is a plot of $F(w)$ in rectangular form (i.e., real–imaginary). Again, as predicted by Theorem 6.1, $F(w)$ is uniformly bounded.

Example 3. A Gaussian function (shown in Figure 6.5) is described mathematically as

$$f(t) = e^{-(t/a)^2}. \tag{6.9}$$

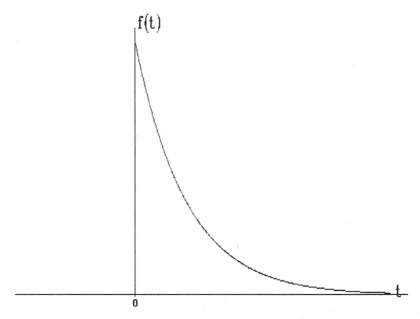

Figure 6.3 One-sided exponential function.

The real number a is a constant greater than zero and is known as the $1/e$ radius of the Gaussian (Note: $f(a) = 1/e$.) This function is very common in statistics and probability theory and is also known as the *normal distribution function*.

We first make note of the fact that

$$\int_R |f(t)|\, dt = \int_R e^{-(t/a)^2}\, dt = (\pi a^2)^{1/2}.$$

Thus $|f(t)| \in L(R)$, and we can use equation (6.1) to calculate its Fourier transform, that is,

$$F(w) = \int_R e^{-(t/a)^2} e^{-2\pi iwt}\, dt,$$

$$= \int_R e^{-(t^2 + 2\pi iwa^2 t)/a^2}\, dt.$$

If we now write $t^2 + 2\pi iwa^2 t$ as $(t + i\pi wa^2)^2 + (\pi wa^2)^2$, then we can re-write the above integral as

$$F(w) = e^{-(\pi wa)^2} \int_R e^{-(t + i\pi w/a)^2}\, dt.$$

Next we make the change of variable $s = (t/a + i\pi w/a)$, which implies

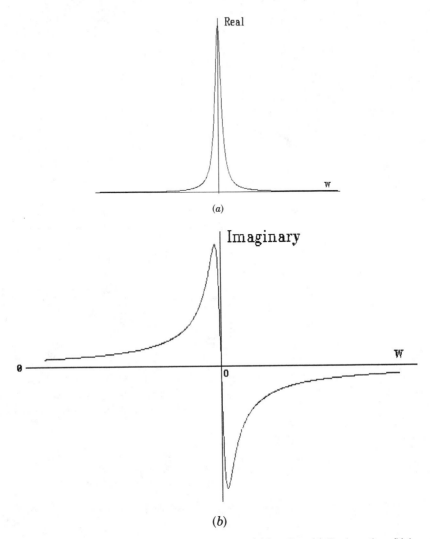

Real

w

(a)

Imaginary

0

W

0

(b)

Figure 6.4 Fourier transform of one-sided exponential function. (a) Real portion; (b) imaginary portion.

$dt = a\,ds$ (note that $dw = 0$). Thus we arrive at

$$F(w) = ae^{-(\pi wa)^2} \int_R e^{-s^2}\,ds.$$

Inasmuch as s is a complex variable, the integral in the above equation must be evaluated using contour integration. When this is done we obtain a value of $(\pi)^{1/2}$. Thus,

$$F(w) = (a^2\pi)^{1/2}e^{-(\pi wa)^2}. \tag{6.10}$$

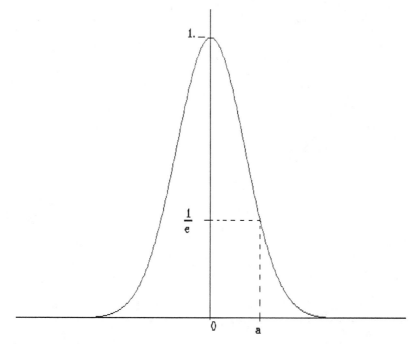

Figure 6.5 The Gaussian function.

We see that $F(w)$ also has a Gaussian profile with $1/e$ radius $1/\pi a$ (see Figure 6.6). This property is known as *self-reciprocity*. When $a = (\pi)^{-1/2}$ we have *complete self-reciprocity*, that is,

$$\mathcal{F}[e^{-\pi t^2}] = e^{-\pi w^2}.$$

PROPERTIES AND BEHAVIOR OF THE FOURIER TRANSFORM

In this section we discuss linearity and continuity of the Fourier transform as well as its behavior at infinity.

Theorem 6.2 (Linearity). If both functions $f(t)$ and $g(t)$ are absolutely integrable and possess Fourier transforms given by $F(w)$ and $G(w)$, respectively, then the function $h(t) = af(t) + bg(t)$ has a Fourier transform given by $aF(w) + bG(w)$.

Proof. By assumption, $|f| \in L(R)$ and $|g| \in L(R)$; consequently $|h| \in L(R)$, where $h = af + bg$. This can be demonstrated by first using Theorem 3.4 to establish $af + bg \in L(R)$ and then using the fact that

$$|h| = |af + bg| \le |af| + |bg|$$

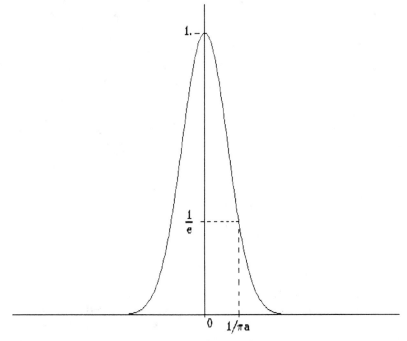

Figure 6.6   Fourier transform of the Gaussian function.

to show that the integral of $|h|$ is bounded over the real line.

Since $|h| \in L(R)$ it makes sense to consider its Fourier transfrom, that is,

$$H(w) = \int_R h(t)e^{-2\pi iwt}\, dt = \int_R (af(t) + bg(t))e^{-2\pi iwt}\, dt.$$

We again use Theorem 3.4 to obtain

$$H(w) = a \int_R f(t)e^{-2\pi iwt}\, dt + b \int_R g(t)e^{-2\pi iwt}\, dt$$

or

$$H(w) = aF(w) + bG(w). \hspace{3cm} \text{Q.E.D.}$$

We now consider the behavior of the Fourier transform $F(w)$ as w approaches infinity.

Theorem 6.3 (Behavior at Infinity). If $|f| \in L(R)$, then

$$\lim_{w \to \infty} F(w) = 0$$

in the sense of the absolute value norm.

Proof. By assumption, $|f| \in L(R)$; thus it makes sense to write

$$F(w) = \int_R f(t)e^{-2\pi i w t}\, dt.$$

Now using Euler's equations we rewrite the above integral as follows:

$$F(w) = \int_R f(t)(\cos(2\pi w t) - i\sin(2\pi w t))\, dt,$$

or

$$\lim_{w \to \infty} F(w) = \lim_{w \to \infty} \int_R f(t)\cos(2\pi w t)\, dt - i \lim_{w \to \infty} \int_R f(t)\sin(2\pi w t)\, dt.$$

However, by definition, $|f| \in L(R)$ implies that $f \in L(R)$, and therefore we can use the Riemann Lebesgue Theorem (Theorem 5.11) to show that both limits on the right-hand side of the above equation equal zero; thus we obtain our desired results.   Q.E.D.

As our last topic in this section we look at continuity of the Fourier transform of a function that is absolutely integrable. First, however, we require the following useful result:

Lemma 6.4. If $|f| \in L(R)$, then given any $\varepsilon > 0$ there exists a real positive number X such that

$$\int_{-\infty}^{X} |f(t)|\, dt < \varepsilon \quad \text{and} \quad \int_{X}^{\infty} |f(t)|\, dt < \varepsilon.$$

Proof. By assumption, $|f| \in L(R)$; therefore there exists a positive number M such that

$$\int_{-\infty}^{\infty} |f(t)|\, dt = M.$$

We next split the above integral and write

$$\int_{-\infty}^{-y} |f(t)|\, dt + \int_{-y}^{y} |f(t)|\, dt + \int_{y}^{\infty} |f(t)|\, dt = M,$$

where y is a positive real number. Now as y increases, the value of

$$\int_{-y}^{y} |f(t)|\, dt$$

can be made arbitrarily close to M since in the limit as y goes to infinity this integral equals M. Therefore, the other two integrals must become arbitrarily small. That is to say, for some finite value of $y = X$ we must have

$$\int_{-\infty}^{-X} |f(t)|\, dt < \varepsilon \quad \text{and} \quad \int_{X}^{\infty} |f(t)|\, dt < \varepsilon.$$ Q.E.D.

This lemmas tells us that if $|f| \in L(R)$, then outside some interval $[-X, X]$ $f(t)$ is arbitrarily close to 0 in the sense of the L^1 norm.* We note that functions that are equal to zero outside of some finite interval are called *functions of bounded support*.

Theorem 6.4 (Continuity). If $|f| \in L(R)$, then the Fourier transform of $f(t)$ is uniformly continuous over the interval R.

Proof. What we wish to prove is: Given any $\varepsilon > 0$, there exists a $\delta > 0$ such that for all $h \in (-\delta, \delta)$ we have

$$|F(w + h) - F(w)| < \varepsilon.$$

We begin by writing

$$F(w + h) - F(w) = \int_R f(t)e^{-2\pi i(w+h)t}\, dt - \int_R f(t)e^{-2\pi iwt}\, dt,$$

$$F(w + h) - F(w) = \int_R f(t)e^{-2\pi iwt}(e^{-2\pi iht} - 1)\, dt.$$

However, it can easily be shown that

$$e^{-2\pi iht} - 1 = -2ie^{-i\pi ht}\sin(\pi ht).$$

Thus,

$$F(w + h) - F(w) = -2i \int_R f(t)e^{-2\pi iwt}e^{-i\pi ht}\sin(\pi ht)\, dt,$$

from which it follows

$$|F(w + h) - F(w)| = 2\left| \int_R f(t)e^{-2\pi iwt}e^{-i\pi ht}\sin(\pi ht)\, dt \right|.$$

Now using Theorem 3.4(d) we have

$$|F(w + h) - F(w)| \leq 2 \int_R |f(t)e^{-2\pi iwt}e^{-i\pi ht}\sin(\pi ht)|\, dt.$$

Next we split the integral as follows:

$$|F(w + h) - F(w)| \leq 2\int_{-\infty}^{-X} |(**)|\, dt + 2\int_{-X}^{X} |(**)|\, dt + 2\int_{X}^{\infty} |(**)|\, dt.$$

Let us now consider each of the above integrals individually. First

$$\int_{-\infty}^{-X} |f(t)e^{-2\pi iwt}e^{-i\pi ht}\sin(\pi ht)|\, dt \leq \int_{-\infty}^{X} |f(t)|\, dt.$$

---

*The L^1 norm of a function f defined on the interval I is defined as the Lebesgue integral of $|f|$ over the interval I.

However, we know from the previous lemma (Lemma 6.4) that, given any $\varepsilon/6 > 0$, we can always find a finite value X_1 such that

$$\int_{-\infty}^{-X_1} |f(t)|\, dt < \frac{\varepsilon}{6};$$

therefore,

$$2\int_{-\infty}^{-X_1} |f(t)e^{-2\pi iwt}e^{-i\pi ht}\sin(\pi ht)|\, dt < \frac{\varepsilon}{3}.$$

Similar reasoning shows that we can always find a finite value of X_2 such that

$$2\int_{X_2}^{\infty} |f(t)e^{-2\pi iwt}e^{-i\pi ht}\sin(\pi ht)|\, dt < \frac{\varepsilon}{3}.$$

If we now choose X to be the largest value of X_1 or X_2 we can establish the fact that the above results are true outside the interval $(-X, X)$.

We now consider the integral

$$2\int_{-X}^{X} |f(t)e^{-2\pi iwt}e^{-i\pi ht}\sin(\pi ht)|\, dt.$$

Over the interval $(-X, X)$ we have

$$|\sin(\pi ht)| \le \pi ht \le \pi hX.$$

Thus,

$$\int_{-X}^{X} |f(t)e^{-2\pi iwt}e^{-i\pi ht}\sin(\pi ht)|\, dt \le \int_{-X}^{X} |f(t)|\pi hX\, dt \le M\pi hX,$$

where M is the bound on the norm of f (by assumption, $|f| \in L(R)$). Therefore if we choose $\delta = \varepsilon/6M\pi X$, we find that for $h \in (-\delta, \delta)$ we have

$$2\int_{-X}^{X} |f(t)e^{-2\pi iwt}e^{-i\pi ht}\sin(\pi ht)|\, dt \le \frac{\varepsilon}{3}.$$

Combining the three previous inequalities, we obtain our desired results, that is

$$|F(w+h) - F(w)| \le \frac{\varepsilon}{3} + \frac{\varepsilon}{3} + \frac{\varepsilon}{3} = \varepsilon. \qquad \text{Q.E.D.}$$

Example 4. For the pulse, the one-sided exponential, and the Gaussian functions, we note that their Fourier transforms, as given by equations (6.6), (6.8), and (6.10), are uniformly continuous over R.

THE SHIFTING THEOREMS

In this section we consider three theorems that permit us to calculate the Fourier transform of the functions $f(t-a)$, $f(t)\exp[2\pi iwa]$, and $f(at)$ from a knowledge of the transform of $f(t)$ itself. We begin with

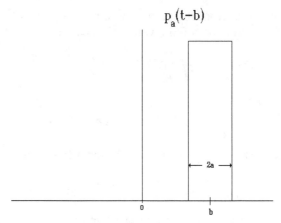

Figure 6.7 Shifted pulse function $p_a(t-b)$.

Theorem 6.5 (First Shifting Theorem). If $|f(t)| \in L(R)$ and has a Fourier transform given by $F(w)$, then the Fourier transform of the function $f(t-a)$ exists and is given as

$$F(w)e^{-2\pi i w a}.$$

Proof. By assumption, $|f(t)| \in L(R)$; consequently $|f(t-a)| \in L(R)$ for all finite values of a. Thus, it makes sense to consider the Fourier transform of $f(t-a)$, that is,

$$\mathcal{F}[f(t-a)] = \int_R f(t-a)e^{-2\pi i w t}\, dt.$$

Making the change of variable $x = t - a$ (which implies $dx = dt$), we obtain

$$\mathcal{F}[f(t-a)] = \int_R f(x)e^{-2\pi i w(x+a)}\, dx,$$

$$\mathcal{F}[f(t-a)] = e^{-2\pi i w a} \int_R f(x)e^{-2\pi i w x}\, dx = F(w)e^{-2\pi i w a}. \qquad \text{Q.E.D.}$$

Example 5. Consider the shifted pulse function (of half-width a) shown in Figure 6.7. We note that this is the same pulse function shown in Figure 6.1, except that it is shifted to the right by an amount b. In other words, we are now dealing with the function $f(t) = p_a(t-b)$, where $p_a(t)$ is described mathematically by equation (6.5). We know from Example 1 that

$$\mathcal{F}[p_a(t)] = 2a\,\text{sinc}(2\pi w a).$$

Consequently, the transform of $p_a(t-b)$ is obtained using Theorem 6.5 as follows:

$$F(w) = 2a\,\text{sinc}(2\pi w a)e^{-2\pi i w b}$$

$$= 2a\,\text{sinc}(2\pi w a)[\cos(2\pi w b) - i\sin(2\pi w b)]. \qquad (6.11)$$

This is shown in Figure 6.8. We note that whereas the Fourier transform of $p(t)$ is a real function, the transform of $p_a(t-b)$ is a complex function with real portion $2a\,\text{sinc}(2\pi wa)[\cos(2\pi wb)]$ and imaginary portion $-2a\,\text{sinc}(2\pi wa)[\sin(2\pi wb)]$.

Theorem 6.6 (Second Shifting Theorem). If $|f(t)| \in L(R)$ has a Fourier transform given by $F(w)$, then the Fourier transform of the function $f(t)\exp[2\pi iat]$ exists and is given by $F(w-a)$.

Proof. We first note that $|f(t)\exp[2\pi iat]| = |f(t)|$; thus, we conclude that $|f(t)\exp[2\pi iat]| \in L(R)$, and it makes sense to consider

$$\mathcal{F}[f(t)e^{-2\pi iat}] = \int_R f(t)e^{-2\pi iat}e^{-2\pi iwt}\,dt$$

$$\mathcal{F}[f(t)e^{-2\pi iat}] = \int_R f(t)e^{-2\pi i(w-a)t}\,dt = F(w-a). \qquad \text{Q.E.D.}$$

It is interesting to note the duality between Theorems 6.5 and 6.6. The first theorem implies that a shift in the time domain results in a complex multiplication in the frequency domain. On the other hand, the second theorem tells us that a complex multiplication in the time domain results in a shift in the frequency domain.

We next consider an immediate consequence of Theorem 6.6.

Corollary 6.6. If $|f(t)| \in L(R)$ has a Fourier transform given by $F(w)$, then we have:

(a) $\mathcal{F}[f(t)\cos(2\pi at)] = [F(w+a) + F(w-a)]/2$ and
(b) $\mathcal{F}[f(t)\sin(2\pi at)] = i[F(w+a) - F(w-a)]/2$.

Proof. If $|f| \in L(R)$, then Theorem 6.6 implies

$$\mathcal{F}[f(t)\cos(2\pi at) + if(t)\sin(2\pi at)] = F(w-a)$$

and

$$\mathcal{F}[f(t)\cos(2\pi at) - if(t)\sin(2\pi at)] = F(w+a).$$

Addition of the previous two equations yields the results of part (a), whereas subtraction yields the results of part (b).   Q.E.D.

Example 6. Let us now determine the Fourier transform of the damped sinusoid, which is shown in Figure 6.9 and described mathematically as

$$g(t) = e^{-\sigma t}\cos(2\pi w_0 t), \qquad t \geq 0. \tag{6.12}$$

From Example 2 we know that the Fourier transform of the one-sided exponential $\exp[-\sigma t]$ is given as

$$\mathcal{F}(w) = \frac{1}{\sigma + 2\pi iw}.$$

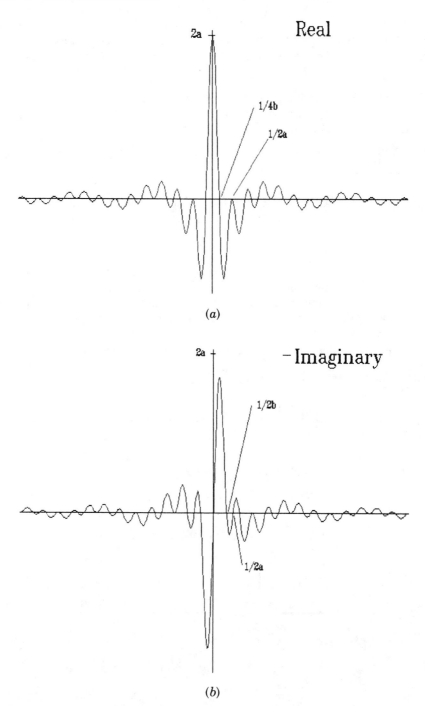

Figure 6.8 Fourier transform of shifted pulse function. (a) Real portion; (b) imaginary portion.

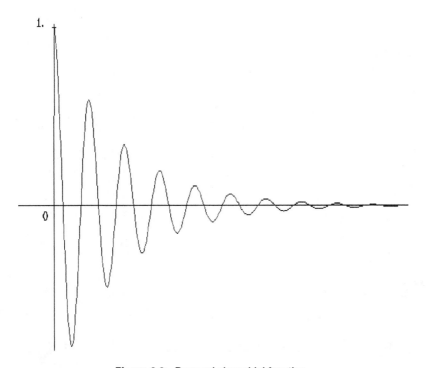

Figure 6.9 Damped sinusoidal function.

Thus, using the results of Corollary 6.6(a), we know

$$\mathcal{F}[e^{-\sigma t}\cos(2\pi w_0 t)] = \frac{1/2}{\sigma + 2\pi i(w - w_0)} + \frac{1/2}{\sigma + 2\pi i(w + w_0)}. \tag{6.13}$$

This transform is shown in Figure 6.10. As can be appreciated by comparing Figures 6.10 and 6.4, the Fourier transform of the damped sinusoid is simply (one-half) the transform of the one-sided exponential function shifted and centered about $+w_0$ and $-w_0$.

We end this section with the following theorem.

Theorem 6.7 (Scale Change). If $|f(t)| \in L(R)$ has a Fourier transform given by $F(w)$, then the Fourier transform of the function $f(at)$ exists and is given by

$$\frac{1}{|a|}F\left(\frac{w}{a}\right),$$

where a is any real number not equal to zero.

Proof. By assumption, $|f(t)| \in L(R)$, which implies $|f(at)| \in L(R)$; thus it makes sense to consider its Fourier transform.

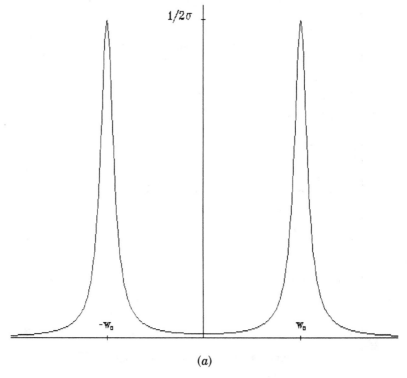

(a)

Figure 6.10 Fourier transform of damped sinusoid. (a) Real portion; (b) imaginary portion.

We first consider the situation where $a > 0$, that is,

$$\mathcal{F}[f(at)] = \int_{-\infty}^{\infty} f(at)e^{-2\pi iwt}\, dt.$$

Making the change of variable $x = at$ (which implies $t = x/a$ and $dt = dx/a$), we obtain

$$\mathcal{F}[f(at)] = \frac{1}{a}\int_{-\infty}^{\infty} f(x)e^{-2\pi iwx}\, dx = \left(\frac{1}{a}\right) F(w).$$

Now for $a < 0$ we make the same change of variable. However, in this case we must also change the sign of the limits of integration, that is,

$$\mathcal{F}[f(at)] = \frac{1}{a}\int_{\infty}^{-\infty} f(x)e^{-2\pi iwx}\, dx.$$

Now interchanging the integration limits (and changing the sign of the integral), we obtain

$$\mathcal{F}[f(at)] = -\frac{1}{a}\int_{-\infty}^{\infty} f(x)e^{-2\pi iwx}\, dx = -\left(\frac{1}{a}\right) F(w).$$

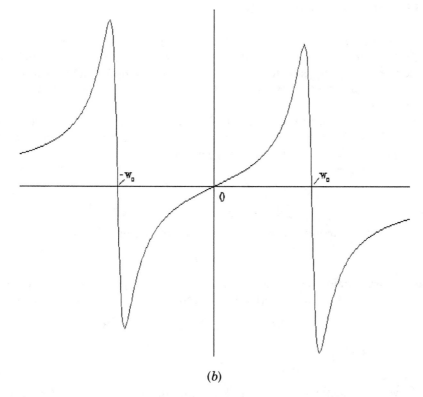

(b)

Figure 6.10   *(Continued)*

Combining the above two cases we obtain our desired results, that is,

$$\mathcal{F}[f(at)] = \left(\frac{1}{|a|}\right) F(w). \qquad \text{Q.E.D.}$$

The obvious consequence of this theorem is presented in the following corollary.

Corollary 6.7. If $|f(t)| \in L(R)$ has a Fourier transform given by $F(w)$, then the Fourier transform of the function $f(-t)$ exists and is given by $F(-w)$.

Let us again pause and consider the dual nature of the Fourier transform that Theorem 6.7 points out. If we stretch the scale in the time domain ($|a| < 1$), then we compress the scale in the frequency domain ($1/|a| > 1$). Similarly, if we compress the scale in the time domain, then we stretch the scale in the frequency domain.

THE DERIVATIVE THEOREMS

In this section we present theorems that deal with the Fourier transform of the derivative and the derivative of the Fourier transform. Simple application of these theorems permits us to obtain the Fourier tranform of additional functions without having to perform the integral operation of equation (6.1).

Theorem 6.8 (Derivative of the Transform). If $|f(t)| \in L(R)$ and $|tf(t)| \in L(R)$, then the Fourier transform of $f(t)$ has a bounded derivative at every point $w \in R$. Furthermore, it is given as

$$\frac{dF(w)}{dw} = -2\pi i \mathcal{F}[tf(t)].$$

Proof. We use Theorem 3.14 to prove our desired results. First let

$$h(t, w) = f(t)e^{-2\pi iwt}.$$

Now, by assumption, $f(t) \in L(R)$, which implies $h(t, w) \in L(R)$. Next, we note that the partial derivative of $h(t, w)$ with respect to w exists and is given by

$$D_2 h(t, w) = -2\pi it f(t)e^{-2\pi iwt}.$$

We also note

$$|D_2 h(t, w)| \leq 2\pi |tf(t)| = G(t).$$

However, by assumption, $G(t) \in L(R)$; therefore we have satisfied the conditions of Theorem 3.14. Consequently,

$$\frac{dF(w)}{dw} = \int_R D_2 h(t, w) dt = -2\pi i \int_R tf(t)e^{-2\pi iwt} dt$$

$$\frac{dF(w)}{dw} = -2\pi i \mathcal{F}[tf(t)]. \qquad \text{Q.E.D.}$$

Repeated application of this theorem yields the following theorem.

Theorem 6.9. If $|f(t)| \in L(R)$ and $|t^n f(t)| \in L(R)$, then the Fourier transform of $f(t)$ possesses all derivatives up to and including the nth. Furthermore, these derivatives are given as

$$\frac{d^n F(w)}{dw^n} = (-2\pi i)^n \mathcal{F}[t^n f(t)].$$

Example 7. Consider the function $f(t) = e^{-at}$ with Fourier transform $F(w) = 1/(a + 2\pi iw)$. By direct application of Theorem 6.8 we find

$$\mathcal{F}[te^{-at}] = \frac{1}{-2\pi i}\frac{dF(W)}{dw} = \frac{-(a + 2\pi iw)^{-2}(2\pi i)}{-2\pi i} = \frac{1}{(a + 2\pi iw)^2}. \qquad (6.14)$$

Example 8. Again let us consider the function $f(t) = e^{-at}$ with Fourier transform $F(w) = (a + 2\pi iw)^{-1}$. We first note that

$$\frac{d^n F(w)}{dw^n} = (n!)(-2\pi i)^n (a + 2\pi iw)^{-(n+1)}.$$

Now applying Theorem 6.9 we have

$$\mathcal{F}[t^n f(t)] = \frac{(n!)(-2\pi i)^n (a + 2\pi iw)^{-(n+1)}}{(-2\pi i)^n} = \frac{n!}{(a + 2\pi iw)^{n+1}}. \qquad (6.15)$$

We next consider the Fourier transform of the derivative of a function. However, we must first establish the following.

Lemma 6.10. If $|f(t)| \in L(R)$ and $|f'(t)| \in L(R)$, then

$$\lim_{t \to \infty} f(t) = 0 \quad \text{and} \quad \lim_{t \to -\infty} f(t) = 0,$$

in the sense of the absolute value norm.

Proof. By assumption, $|f'(t)| \in L(R)$; therefore, given any $\varepsilon > 0$, there exists a finite real number X such that

$$\int_X^\infty |f'(t)| \, dt < \varepsilon.$$

Thus, we can establish

$$|f(\infty) - f(X)| = \left| \int_X^\infty f'(t) \, dt \right| \le \int_X^\infty |f'(t)| \, dt < \varepsilon.$$

Thus we see that $f(t)$ must approach a definite limit as t approaches ∞. We now determine what that limit is. Since $|f| \in L(R)$, given any $\varepsilon > 0$, there exists a positive real number Y such that

$$\int_Y^\infty |f(t)| \, dt < \varepsilon,$$

and we see that over the interval (Y, ∞), $|f(t)|$ is arbitrarily close to 0, or, in other words,

$$\lim_{t \to \infty} f(t) = 0.$$

Similar reasoning yields

$$\lim_{t \to -\infty} f(t) = 0. \qquad \text{Q.E.D.}$$

With this preliminary result in hand, we present the following theorem.

Theorem 6.10 (Transform of the Derivative). If $F(w)$ is the Fourier transform of the function $|f(t)| \in L(R)$ and $|f'(t)| \in L(R)$, then $f'(t)$ possesses a Fourier transform that is given as $2\pi iw F(w)$.

Proof. Since $|f'(t)| \in L(R)$, it makes sense to consider its Fourier transform, that is,

$$\mathcal{F}[f'(t)] = \int_R f'(t)e^{-2\pi iwt}\, dt.$$

Using integration by parts in the above equation, we obtain

$$\mathcal{F}[f'(t)] = f(t)e^{-2\pi iwt}\Big|_{-\infty}^{\infty} + 2\pi iw \int_R f(t)e^{-2\pi iwt}\, dt.$$

Now from the previous lemma (Lemma 6.10) we know $f(\infty) = f(-\infty) = 0$ and we obtain our desired result, that is,

$$\mathcal{F}[f'(t)] = 2\pi iw \int_R f(t)e^{-2\pi iwt}\, dt. \qquad \text{Q.E.D.}$$

Theorem 6.10 can be generalized by repeated application to yield the following theorem.

Theorem 6.11. If $F(w)$ is the Fourier transform of the function $|f(t)| \in L(R)$ and $|f^{[n]}(t)| \in L(R)$, then $f^{[n]}(t)$ possesses a Fourier transform that is given as $(2\pi iw)^n F(w)$.

We can combine Theorems 6.11 and 6.3 to obtain the following result describing additional behavior of the Fourier transform $F(w)$ at infinity.

Theorem 6.12. If $|f(t)| \in L(R)$ and $|f^{[n]}(t)| \in L(R)$, then, at infinity, $F(w)$ goes to zero faster than w^n goes to infinity, that is,

$$\lim_{w \to \infty} (2\pi iw)^n F(w) = 0.$$

Let us again consider the duality of the Fourier transform. Theorem 6.8 tells us that the derivative of the Fourier transform is equal to the Fourier transform of the original function multiplied by $-2\pi it$. On the other hand, Theorem 6.11 says that the Fourier transform of the derivative of a function is equal to the transform of the original function multiplied by $2\pi iw$. In other words, a derivative in the time domain corresponds to a multiplication in the frequency domain, while a derivative in the frequency domain corresponds to a multiplication in the time domain.

Example 9. Consider the Gaussian function of equation (6.9) with $a = 1$, that is,

$$f(t) = e^{-t^2}.$$

The transform of this function (see Example 3, equation (6.10)) is

$$F(w) = \pi^{1/2}e^{-(\pi w)^2}.$$

We note that the derivative of $f(t)$ in the above equation is

$$g(t) = f'(t) = -2te^{-t^2}. \tag{6.16}$$

Thus, applying Theorem 6.10 we find

$$G(w) = 2\pi^{3/2}iwe^{-(\pi w)^2},$$

or using the Linearity Theorem (Theorem 6.2) we have

$$\mathcal{F}[te^{-t^2}] = -i\pi^{3/2}we^{-(\pi w)^2}. \tag{6.17}$$

TRANSFORM OF A TRANSFORM

Several times in this chapter we have noted the dual nature of the Fourier transform. In particular, we noted the duality expressed by the shifting and derivative theorems. Actually, if we examine equations (6.1) and (6.2) and note their similar form, this should not be too surprising. In this section we again explore the similar nature of the transform and its inverse. We do so by considering the Fourier transform of a Fourier transform.

Theorem 6.13 (Transform of a Transform). If $f(t)$ and $F(w)$ are Fourier transform pairs and both are absolutely integrable (i.e., $|f(t)| \in L(R)$ and $|F(w)| \in L(R)$), then

$$\mathcal{F}[\mathcal{F}[f(t)]] = f(-t).$$

Proof. By assumption, $|f| \in L(R)$; thus it makes sense to write

$$\mathcal{F}[f(t)] = F(w).$$

By the same token, $|F| \in L(R)$; therefore it also makes sense to consider

$$\mathcal{F}[\mathcal{F}[f(t)]] = \mathcal{F}[F(w)] = \int_R F(w)e^{-2\pi iwt} \, dw = \int_R F(w)e^{2\pi iw(-t)} \, dw.$$

Now by comparing the above equation to the definition of the inverse Fourier transform of equation (6.2), we see that it is in fact an inverse transform equal to $f(-t)$; this establishes our desired results, that is,

$$\mathcal{F}[\mathcal{F}[f(t)]] = f(-t). \qquad \text{Q.E.D.}$$

This theorem illustrates the fact that if we perform the Fourier transform operation of equation (6.1) twice on a function $f(t)$, we obtain the function rotated about the vertical axis, that is, $f(-t)$.

SYMMETRY CONSIDERATIONS

Up to this point we have considered the Fourier transform of absolutely integrable functions that were general in nature. In other words, there were no restrictions placed on the function as to their symmetry or other behavior. In this section we consider the nature of the Fourier transform of (absolutely integrable) functions that are limited in some way such as, for example, always being real valued or even.

We first consider the situation when the function being transformed is real. In actual practice, many, if not most, of the functions to be transformed are real. The most general form of a function is one that is complex with both a real and an imaginary portion. When we say a function is *real* we mean that it has no imaginary portion and it is equal to its complex conjugate,[†] that is, $f^*(t) = f(t)$.

Theorem 6.14 (Transform of a Real Function). If f is a real function that is absolutely integrable, then its Fourier transform $F(w)$ has the following property:

$$F(-w) = F^*(w).$$

Proof. Since $|f| \in L(R)$ we can write

$$F(w) = \int_R f(t)e^{-2\pi i w t}\, dt.$$

Now taking the complex conjugate of both sides of the above equation, we find

$$F^*(w) = \int_R f^*(t)e^{2\pi i w t}\, dt = \int_R f(t)e^{2\pi i w t}\, dt$$

or

$$F^*(w) = \int_R f(t)e^{-2\pi i (-w)t}\, dt = F(-w). \qquad \text{Q.E.D.}$$

Example 10. In Example 2 we determined the Fourier transform of the real function $f(t) = \exp[-at]$ to be given as $F(w) = 1/(a + 2\pi i w)$. For this function we note $F^*(w) = 1/(a - 2\pi i w) = F(-w)$.

For any absolutely integrable function, we have

$$F(w) = \int_R f(t)e^{-2\pi i w t}\, dt = \int_R f(t)[\cos(2\pi w t) - i\sin(2\pi w t)]\, dt$$

or

$$F(w) = \int_R f(t)\cos(2\pi w t)\, dt - i\int_R f(t)\sin(2\pi w t)\, dt. \qquad (6.18)$$

[†]The complex conjugate of a complex function $f(t) = R(t) + iI(t)$ is denoted as $f^*(t)$ and is defined as $f^*(t) = R(t) - iI(t)$. In other words, we change the sign of the imaginary portion.

Now when $f(t)$ is a real function we have the real and imaginary portions of its Fourier transform given as

$$\text{Re}\{F(w)\} = \int_R f(t)\cos(2\pi wt)\,dt, \tag{6.19}$$

$$\text{Im}\{F(w)\} = -\int_R f(t)\sin(2\pi wt)\,dt. \tag{6.20}$$

Example 11. Let us now determine the real portion of the Fourier transform of the one-sided exponential function of equation (6.7). We accomplish this by applying equation (6.19), that is,

$$\text{Re}\{F(w)\} = \int_R f(t)\cos(2\pi wt)\,dt = \int_R e^{-at}\cos(2\pi wt)\,dt.$$

Using integration by parts (or an integral table), we obtain

$$\text{Re}\{F(w)\} = \frac{e^{-at}}{a^2 + (2\pi w)^2}[2\pi w\sin(2\pi wt) - a\cos(2\pi wt)]\Big|_0^\infty$$

or

$$\text{Re}\{F(w)\} = \frac{a}{a^2 + (2\pi w)^2},$$

which agrees with the results of equation (6.8).

Returning to equation (6.18) for the general case when $f(t)$ is a complex function (i.e., $f(t) = R(t) + iI(t)$), we obtain

$$F(w) = \int_R [R(t) + iI(t)]\cos(2\pi wt)\,dt - i\int_R [R(t) + iI(t)]\sin(2\pi wt)\,dt$$

$$= \int_R [R(t)\cos(2\pi wt) + I(t)\sin(2\pi wt)]\,dt$$

$$+ i\int_R [I(t)\cos(2\pi wt) - R(t)\sin(2\pi wt)]\,dt.$$

Consequently,

$$\text{Re}\{F(w)\} = \int_R [R(t)\cos(2\pi wt) + I(t)\sin(2\pi wt)]\,dt, \tag{6.21}$$

$$\text{Im}\{F(w)\} = \int_R [I(t)\cos(2\pi wt) - R(t)\sin(2\pi wt)]\,dt. \tag{6.22}$$

We first note that when the function is real (i.e., $f(t) = R(t)$ and $I(t) = 0$), then the above equations reduce to equations (6.19) and (6.20). On the other hand, when the function is imaginary (i.e., $f(t) = I(t)$ and $R(t) = 0$), then we obtain

$$\text{Re}\{F(w)\} = \int_R f(t)\sin(2\pi wt)\,dt, \qquad (6.23)$$

$$\text{Im}\{F(w)\} = \int_R f(t)\cos(2\pi wt)\,dt. \qquad (6.24)$$

Let us now consider the situation when the function $f(t)$ is even (i.e., $f(-t) = f(t)$). We begin with the definition of the Fourier transform:

$$F(w) = \int_R f(t)e^{-2\pi iwt}\,dt.$$

We next split the integral as follows:

$$F(w) = \int_{-\infty}^{0} f(t)e^{-2\pi iwt}\,dt + \int_{0}^{\infty} f(t)e^{-2\pi iwt}\,dt.$$

Now in the first integral we substitute $-t$ for t to obtain

$$F(w) = -\int_{\infty}^{0} f(-t)e^{2\pi iwt}\,dt + \int_{0}^{\infty} f(t)e^{-2\pi iwt}\,dt.$$

Finally, we interchange the limits on the first integral (which also changes the sign of the integral) and make use of the fact that $f(-t) = f(t)$. This results in

$$F(w) = \int_{0}^{\infty} f(t)e^{2\pi iwt}\,dt + \int_{0}^{\infty} f(t)e^{-2\pi iwt}\,dt = \int_{0}^{\infty} f(t)[e^{2\pi iwt} + e^{-2\pi iwt}]\,dt$$

or

$$F(w) = 2\int_{0}^{\infty} f(t)\cos(2\pi wt)\,dt.$$

Similar reasoning shows that the Fourier transform of an odd function (i.e., $f(-t) = -f(t)$) is given as

$$F(w) = -2i\int_{0}^{\infty} f(t)\sin(2\pi wt)\,dt.$$

We summarize the previous development as the following.

Theorem 6.15. Given $|f(t)| \in L(R)$, the following conditions hold:

(a) If f is an even function, then

$$F(w) = 2\int_{0}^{\infty} f(t)\cos(2\pi wt)\,dt.$$

(b) If f is an odd function, then

$$F(w) = -2i\int_{0}^{\infty} f(t)\sin(2\pi wt)\,dt.$$

Example 12. We now calculate the Fourier transform of the pulse function of equation (6.5). Since the pulse function $p_a(t)$ is an even function, we use Theorem 6.15(a), that is,

$$F(w) = 2 \int_0^\infty p_a(t) \cos(2\pi wt) \, dt = 2 \int_0^a \cos(2\pi wt) \, dt,$$

$$F(w) = \frac{2 \sin(2\pi wt)}{2\pi w} \bigg|_0^a = \frac{2 \sin(2\pi wa)}{2\pi w} = 2a \operatorname{sinc}(2\pi wa),$$

which agrees with the previously obtained result of equation (6.6).

Finally, we look at the symmetry of the Fourier transform of a function that is a combination of odd and even as well as real and imaginary.

Theorem 6.16. Assume that the function $f(t)$ is absolutely integrable with a Fourier transform $F(w)$, such that:

(a) If $f(t)$ is real and even, then $F(w)$ is real and even.
(b) If $f(t)$ is real and odd, then $F(w)$ is imaginary and odd.
(c) If $f(t)$ is imaginary and even, then $F(w)$ is imaginary and even.
(d) If $f(t)$ is imaginary and odd, then $F(w)$ is real and odd.

Proof. (a) By assumption, $f(t)$ is real and even; this fact implies that $f^*(t) = f(t)$ and $f(-t) = f(t)$ or $f^*(-t) = f(t)$. Also, since $|f| \in L(R)$ it makes sense to write

$$F^*(w) = \int_R f^*(t) e^{2\pi iwt} \, dt.$$

Now substitution of $-t$ for t in the above equation yields

$$F^*(w) = \int_R f^*(-t) e^{-2\pi iwt} \, dt = \int_R f(t) e^{-2\pi iwt} \, dt = F(w);$$

thus we see $F(w)$ is a real function. Next we demonstrate that $F(w)$ is also even. Again, by definition we have

$$F(-w) = \int_R f(t) e^{2\pi iwt} \, dt.$$

Substitution of $-t$ for t in the above equation yields our desired results, that is,

$$F(-w) = \int_R f(-t) e^{-2\pi iwt} \, dt = \int_R f(t) e^{-2\pi iwt} \, dt = F(w).$$

The proofs of parts (b)–(d) are similar and are left as an exercise.   Q.E.D.

UNIQUENESS AND RECIPROCITY

Up to this point we have only been concerned with the existence of the Fourier transform of absolutely integrable functions. We have really said nothing about the existence of the inverse Fourier transform. We now consider this question and examine the conditions under which we can obtain the function $f(t)$ from its transform $F(w)$.

We begin by considering the definition of the Fourier transform as given by equation (6.1), that is,

$$F(w) = \int_R f(t)e^{-2\pi iwt}\, dt.$$

Recall that this integral is performed in the Lebesgue sense, and consequently it is independent of the behavior of the function $f(t)$ on a set of measure 0. That is to say, $F(w)$ is unchanged by changing the value of $f(t)$ at a number of individual points (a subset of R of measure 0). For example, we have already demonstrated that the Fourier transform of the Gaussian function

$$f(t) = e^{-(t/a)^2}$$

is given as

$$F(w) = (a^2\pi)^{1/2}e^{-(\pi wa)^2}.$$

Now, let's consider the function $g(t)$, defined as

$$g(t) = \begin{cases} 0, & \text{if } t \text{ is an integer value,} \\ e^{-(t/a)^2}, & \text{otherwise.} \end{cases}$$

This function is shown in Figure 6.11. We note that $g(t)$ differs from the Gaussian function (see Figure 6.5) only at integer values of t, which form a set of measure 0. Therefore,

$$G(w) = \int_R g(t)e^{-2\pi iwt}\, dt = \int_R f(t)e^{-2\pi iwt}\, dt = F(w).$$

Let's consider this situation more closely. We have two different functions $f(t)$ and $g(t)$ with the same Fourier transform. We see, therefore, that only conditions of continuity placed on a function guarantee that its transform is unique. In summary, we have: If $|f(t)| \in L(R)$, then we can guarantee that its Fourier transform $F(w)$ will exist, is uniformly continuous, and is bounded, but it does not guarantee that $F(w)$ is unique.

We now examine this problem of obtaining the function $f(t)$ from its Fourier transform. Obviously the previous demonstration implies that $f(t)$ cannot always be uniquely determined from $F(w)$. To study this more closely, let us begin with the following integral:

$$h(t) = \int_R F(w)e^{2\pi iwt}\, dw. \tag{6.25}$$

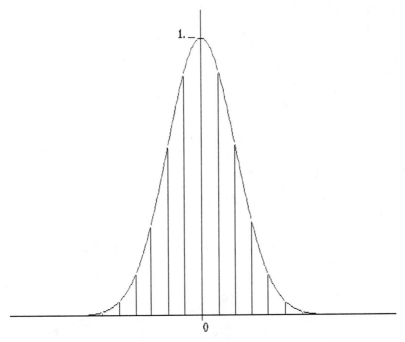

Figure 6.11 Gaussian function with "holes" at integer values of [t].

Our main concern here is: Does $h(t)$ exist and, if it does, is it equal to $f(t)$ a.e. on R? We first note that if $|F(w)| \in L(R)$, then the integral of equation (6.25) does indeed exist and we have

$$h(t) = \lim_{a \to -\infty} \int_a^c F(w)e^{2\pi i w t}\, dw + \lim_{b \to \infty} \int_c^b F(w)e^{2\pi i w t}\, dw.$$

We note that $|F| \in L(R)$ implies that the upper and lower limits in the above equation may approach infinity independently of each other. In this case, however, we require that the limits approach infinity while remaining equal. In other words, we consider the principal value (PV) of the above integral, and we rewrite equation (6.25) as

$$h(t) = \lim_{a \to \infty} g(t,a) = \lim_{a \to \infty} \int_{-a}^a F(w)e^{2\pi i w t}\, dw. \tag{6.26}$$

We first use equation (6.2) for $F(w)$ and substitute this into the above equation to obtain

$$g(t,a) = \int_{-a}^a e^{2\pi i w t} \int_{-\infty}^{\infty} f(x)e^{-2\pi i w x}\, dx\, dw.$$

Now since $\exp[2\pi iwt] \in L([-a,a])$ and $f(t) \in L(R)$, we can use Theorem 3.15 to interchange the order of integration to obtain

$$f(t,a) = \int_{-\infty}^{\infty} f(t) \int_{-a}^{a} e^{-2\pi iw(t-x)} dw\,dx.$$

The evaluation of the second integral is straightforward and we find

$$g(t,a) = 2a \int_{-\infty}^{\infty} f(t)\,\mathrm{sinc}(2\pi a(t-x))\,dx.$$

Substitution of y for $t-x$ yields

$$g(x,a) = 2a \int_{-\infty}^{\infty} f(x+y)\,\mathrm{sinc}(2\pi ay)\,dy.$$

Since $\mathrm{sinc}(y)$ is an even function, we are able to split the integral and perform the necessary manipulations to obtain

$$g(t,a) = 2a \int_{0}^{\infty} [f(t+y)+f(t-y)]\,\mathrm{sinc}(2\pi ay)\,dy.$$

(Note: For notational convenience we have substituted t for x in the above equation). We now return to equation (6.26) and consider the limit of $g(t,a)$ as $a \to \infty$. However, we first split the integral as follows:

$$\lim_{a\to\infty} g(t,a) = \lim_{a\to\infty} 2a \int_{0}^{\delta} [f(t+y)+f(t-y)]\,\mathrm{sinc}(2\pi ay)\,dy$$

$$+ \lim_{a\to\infty} 2a \int_{\delta}^{\infty} [f(t+y)+f(t-y)]\,\mathrm{sinc}(2\pi ay)\,dy.$$

Now, since $|f(y)| \in L(R)$, both $f(t+y)/y$ and $f(t-y)/y$ are also absolutely integrable over the interval (δ,∞). Therefore, we can use the Riemann Lebesgue theorem (Theorem 5.11) to conclude that the second integral in the above equation vanishes. We are then left with

$$\lim_{a\to\infty} g(t,a) = \lim_{a\to\infty} 2a \int_{0}^{\delta} [f(t+y)+f(t-y)]\,\mathrm{sinc}(2\pi ay)\,dy.$$

We note, however, that (1) over the interval of integration (i.e., $[0,\delta]$) y is restricted to an arbitrarily small region of the origin and (2) over this region the function $f(t)$ is essentially a constant. Therefore, $f(t+y)$ and $f(t-y)$ are also essentially constants equal to $f(t+)$ and $f(t-)$, respectively, and can be removed from under the integral, that is,

$$\lim_{a\to\infty} g(t,a) = 2a[f(t+)+f(t-)] \lim_{a\to\infty} \int_{0}^{\delta} \mathrm{sinc}(2\pi ay)\,dy.$$

If we now make the change of variable $\tau = 2\pi a y$ (which implies $d\tau = dt/2\pi a$), we obtain

$$\lim_{a \to \infty} g(t,a) = \frac{[f(t+) + f(t-)]}{\pi} \lim_{a \to \infty} \int_0^{2\pi a \delta} \text{sinc}(\tau) d\tau.$$

Finally, as we saw in Chapter 5, the above integral is equal to $\pi/2$. When this is used in the above equation, we obtain

$$h(t) = \frac{f(t+) + f(t-)}{2}. \tag{6.27}$$

Obviously, when $f(t)$ is continuous, $f(t+) = f(t-) = f(t)$ and we see $h(t) = f(t)$. Thus, as we previously stated, when the function $f(t)$ is continuous (and $|F(w)| \in L(R)$) the function $f(t)$ can be uniquely determined from its Fourier transform by use of equation (6.2). When $f(t)$ has a discontinuity at $t = a$, equation (6.27) tells us that $h(t)$ will equal the average value of $f(a+)$ and $f(a-)$. This is very similar to the convergence behavior of the Fourier series at a discontinuity.

We end this section with the interesting observation that when both $f(t)$ and $F(w)$ are absolutely integrable, there exists a perfect reciprocity between them in terms of their Fourier transform pairs. This serves to reinforce the duality concept discussed throughout this chapter.

CONVOLUTION OF TWO FUNCTIONS

The convolution of two functions $f(t)$ and $g(t)$, denoted as $f(t)*g(t)$, is mathematically defined as follows:

$$f(t)*g(t) = \int_R f(x)g(t-x)dx. \tag{6.28}$$

Convolution is a law of composition that combines two functions to yield a third. Inasmuch as convolution shares many similar properties (associativity, commutivity, and distributivity) with the simple product of two functions, it is sometimes called the *convolution product* of two functions. However, rather than a simple point-by-point multiplication, the convolution product of two functions is carried out as per the operation given in equation (6.28).

The convolution of two functions is commutative, that is,

$$f(t)*g(t) = g(t)*f(t).$$

This is demonstrated as follows:

$$f(t)*g(t) = \int_{-\infty}^{\infty} f(x)g(t-x)dx.$$

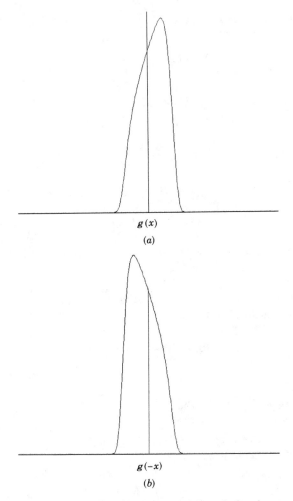

Figure 6.12   Graphical representation of $g(t - x)$.

If we now let $t - x = y$ (which implies $x = t - y$ and $dx = -dy$), then we have

$$f(t) * g(t) = -\int_{\infty}^{-\infty} f(t - y)g(y)\,dy,$$

$$f(t) * g(t) = \int_{-\infty}^{\infty} g(y)f(t - y)\,dy = g(t) * f(t).$$

Using similar mathematical logic, we can also demonstrate that convolution is associative, that is,

$$f(t) * [g(t) * h(t)] = [f(t) * g(t)] * h(t).$$

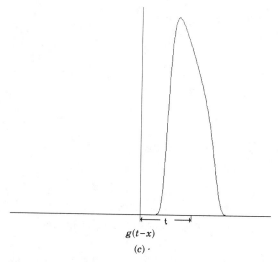

$g(t-x)$

(c) ·

Figure 6.12 *(Continued)*

Finally, convolution is distributive with respect to addition, that is,

$$f(t)*[g(t)+h(t)] = f(t)*g(t)+f(t)*h(t),$$

and

$$[g(t)+h(t)]*f(t) = g(t)*f(t)+h(t)*f(t).$$

We demonstrate this as follows:

$$f(t)*[g(t)+h(t)] = \int_R f(t-x)[g(x)+h(x)]dx.$$

Now using the linearity property of the integral, we have

$$f(t)*[g(t)+h(t)] = \int_R f(t-x)g(x)dx + \int_R f(t-x)h(x)dx,$$

$$f(t)*[g(t)+h(t)] = f(t)*g(t)+f(t)*h(t).$$

The second half of this proposition is demonstrated in a similar way.

In an attempt to provide insight into the nature of the convolution product, we now present a graphical interpretation of the convolution integral. From equation (6.28) we see that the convolution of f and g is given as the integral of the product of two functions $f(x)$ and $g(t-x)$. Let us first consider a graphical representation of $g(t-x)$. Shown in Figure 6.12(a) is the function $g(x)$. In Figure 6.12(b) we have rotated this function about the origin to obtain $g(-x)$, and in Figure 6.12(c) we have displaced this rotated function to the right (for positive values of t) by an amount t to obtain $g(t-x)$.

In Figure 6.13 we show this rotated and displaced function $g(t-x)$ along with the other function $f(x)$. The product of these two functions is shown as

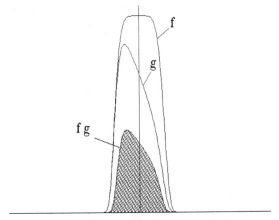

Figure 6.13   Product of f(x) and g(t − x).

the third curve in the figure. The convolution of f and g (for this particular value of t) is the area under this product curve (see equation (6.28)). Figure 6.14 shows this convolution product for four distinct values of t. As can be appreciated from these figures, the convolution product $h(t) = f(t)*g(t)$ for all values of t on the real line is obtained by sweeping this rotated function from $-\infty$ to ∞. This is best illustrated with examples.

Example 13. We first consider the convolution of two equal pulse functions of half-width a as shown in Figure 6.15. Mathematically, the convolution product is given as

$$h(t) = \int_{-\infty}^{\infty} p_a(x)p_a(t-x)\,dx.$$

The task of evaluating this integral lies in determining the proper limits of integration. As can be seen in Figure 6.15, we have four possible situations as the first pulse is rotated and shifted. At first the two do not overlap ($t < -2a$) and the resulting product is zero. When $t = -2a$ they begin to overlap and continue to do so until $t = +2a$. The amount of overlap that begins at $t = -2a$ increases linearly until $t = 0$, where it reaches a maximum (area = $2a$) and then decreases linearly until $t = 2a$. Finally, when $t > 2a$ there is no overlap and again the resulting product is zero. From these comments we can see that the convolution of two pulse functions must be a triangle function of half-width $2a$ and central value of $2a$.

Mathematically, the four separate domains are summarized as:

1. $t < -2a$,
2. $-2a \le t < 0$,
3. $0 \le t \le 2a$,
4. $t > 2a$.

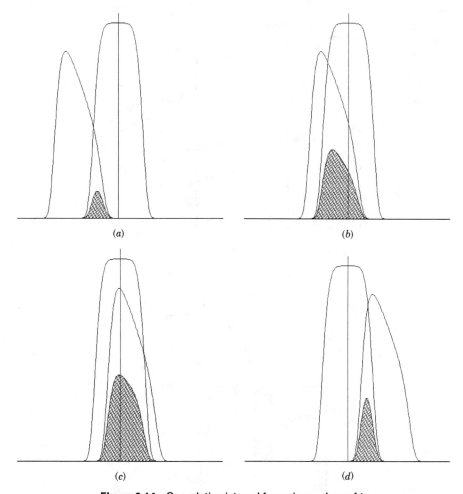

Figure 6.14 Convolution integral for various values of *t*.

As mentioned above, for cases 1 and 4 the two functions do not overlap and the product (and consequently the convolution) of these two functions is zero.

For case 2 (Figure 6.15(b)) the integral becomes

$$h(t) = \int_{-a}^{t+a} dx = t + 2a.$$

Similarly, for case 3 (Figure 6.15(c)) we have

$$h(t) = \int_{t-a}^{a} dx = 2a - t.$$

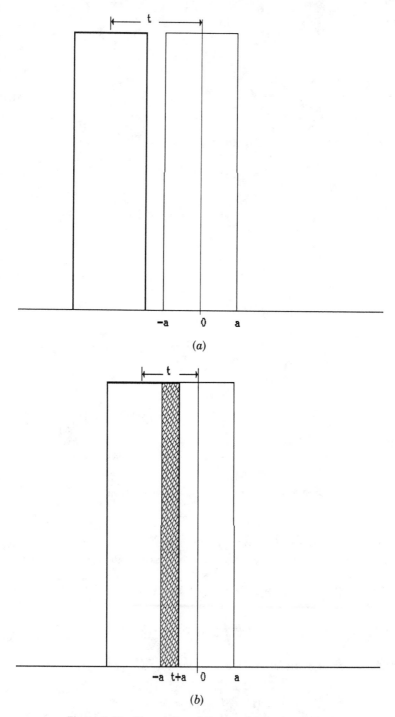

Figure 6.15 Convolution of two equal pulse functions.

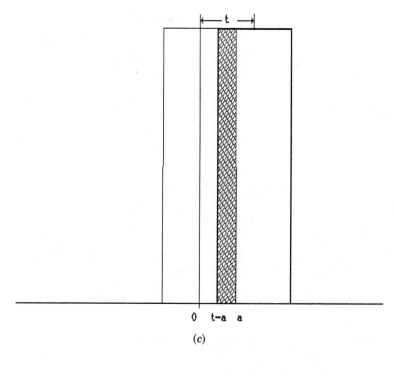

0   t-a   a

(c)

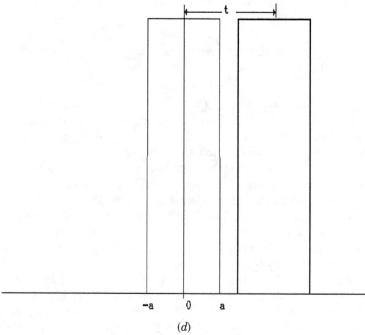

-a   0   a

(d)

Figure 6.15 (Continued)

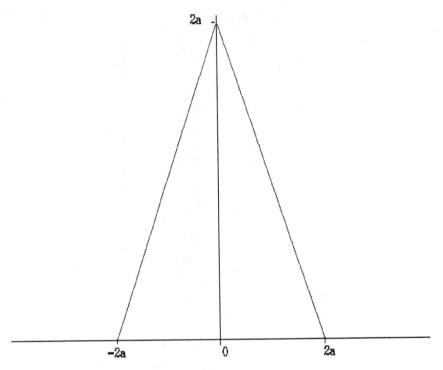

Figure 6.16   Triangle function resulting from convolution of two equal pulse functions.

Thus, again we see that the result is a triangle function of half-width $2a$, which is denoted as $T_{2a}(t)$ and shown in Figure 6.16.

Example 14. As a second example, let us consider the convolution of two pulses of half-widths a and b. For the sake of discussion let us assume that $b > a$. This convolution product is illustrated graphically in Figure 6.17, in which we have rotated and shifted the narrower pulse $p_a(t)$. In this situation we have five different domains to consider. First, for t less than $-(a+b)$ or greater than $(a+b)$ there is no overlap, and consequently the convolution product is zero. When t reaches $-(a+b)$, $p_a(t)$ begins to overlap the wider pulse $p_b(t)$ and the amount of overlap increases linearly until $t = -(b-a)$. Next, from $t = -(b-a)$ to $+(b-a)$ the pulse $p_a(t)$ remains completely within envelope of the wider pulse $p_b(t)$, and the area of overlap remains constant and equal to $2a$. Finally from $t = (b-a)$ to $(a+b)$ the amount of overlap decreases linearly from $2a$ to zero. From these comments we determine that this convolution is equal to the pyramid-type function shown in Figure 6.18.

We next consider the Fourier transform of the convolution product. From a practical, or applications, point of view this is perhaps the most important result of Fourier analysis.

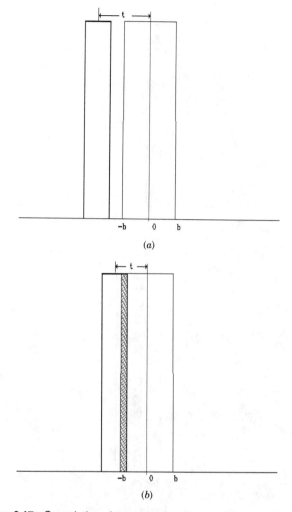

Figure 6.17 Convolution of two pulse functions of different half-widths.

Theorem 6.17 (Convolution). If $|f(t)| \in L(R)$ and $|G(w)| \in L(R)$, then $f(t)$ has a bounded transform $F(w)$ and $G(w)$ has a bounded inverse transform $g(t)$. Furthermore, we have $F(w)G(w) \in L(R)$ and

$$\mathcal{F}[f(t)*g(t)] = F(w)G(w).$$

Proof. We first note that Theorem 6.1 can be used to guarantee that $F(w)$ and $g(t)$ are uniformly bounded. Now, by definition we have

$$h(t) = f(t)*g(t) = \int_R f(x)g(t-x)\,dx.$$

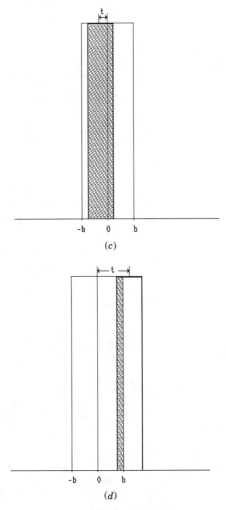

Figure 6.17   *(Continued)*

Using the first shifting theorem (Theorem 6.5), we know

$$\mathcal{F}[g(t-x)] = G(w)e^{-2\pi iwx}$$

or

$$g(t-x) = \int_R G(w)e^{-2\pi iwx}e^{2\pi iwt}\,dw.$$

When this is substituted into the previous equation, we obtain

$$h(t) = \int_R f(x)\int_R G(w)e^{2\pi iw(t-x)}\,dw\,dx.$$

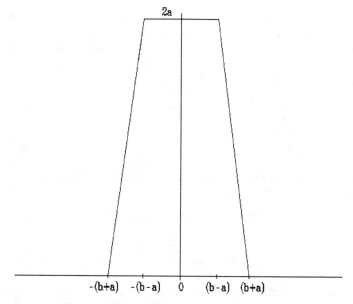

Figure 6.18 Pyramid function resulting from the convolution of two unequal pulse functions.

However, since both f and G are Lebesgue integrable, we can interchange the order of integration to obtain

$$h(t) = \int_R G(w)e^{2\pi iwt}\, dw \int_R f(t)e^{-2\pi iwx}\, dx,$$

or, using equation (6.1), we find

$$h(t) = \int_R G(w)F(w)e^{2\pi iwt}\, dw.$$

The fact that $F(w)$ is bounded and $G(w)$ is absolutely integrable guarantees that the above integral exists. If we now compare this to equation (6.2), we see that $h(t)$ is the inverse Fourier transform of $F(w)G(w)$, or, in other words,

$$\mathcal{F}[h(t)] = F(w)G(w). \qquad\qquad \text{Q.E.D.}$$

Note that this theorem does not require that both $f(t)$ and $g(t)$ be absolutely integrable but requires, instead, that $f(t)$ and $G(w)$ be absolutely integrable. Since convolution is commutative, we could just as well have required that $g(t)$ and $F(w)$ be absolutely integrable and, in doing so, obtain the same results.

The convolution theorem turns out to be a surprisingly nice result. It tells us that the Fourier transform of the convolution product of two functions is simply the multiplication product of the individual transforms.

Example 15. In Example 13 we demonstrated that the convolution of two equal pulses $p_a(t)$ is equal to the triangle function $T_{2a}(t)$ shown in Figure 6.16. Also, from the results of Example 1 we know that the Fourier transform of a pulse of half-width a is given as $F(w) = 2a \operatorname{sinc}(2\pi wa)$. Therefore, applying the convolution theorem (Theorem 6.17), we have

$$\mathcal{F}[T_{2a}(t)] = \mathcal{F}[p_a(t) * p_a(t)] = [2a \operatorname{sinc}(2\pi wa)][2a \operatorname{sinc}(2\pi wa)],$$

$$\mathcal{F}[T_{2a}(t)] = 4a^2 \operatorname{sinc}^2(2\pi wa).$$

Example 16. In Example 14 we showed that the pyramid function shown in Figure 6.18 is given as the convolution of $p_a(t)$ and $p_b(t)$. Consequently, the convolution theorem (Theorem 6.17) implies

$$\mathcal{F}[\text{pyramid}] = \mathcal{F}[p_a(t) * p_b(t)] = 4ab \operatorname{sinc}(2\pi wa) \operatorname{sinc}(2\pi wb).$$

One of the real strengths of the convolution theorem is that it provides a convenient method of forming the convolution product of two functions. The method is to multiply the Fourier transforms of the individual functions and then inverse transform the resulting product. This method is shown to be mathematically correct as follows:

$$h(t) = f(t) * g(t);$$

thus,

$$H(w) = F(w)G(w) \quad \text{and} \quad h(t) = \mathcal{F}^{-1}[H(w)] = \mathcal{F}^{-1}[F(w)G(w)].$$

This technique is particularly attractive to use when the transforms are performed digitally using the fast Fourier transform algorithm (see Chapter 9). We now, however, analytically illustrate the technique in the following.

Example 17. Let us perform the convolution product of two Gaussian functions of $1/e$ radii a and b. That is to say, we wish to calculate the following convolution product:

$$h(t) = \exp[-(t/a)^2] * \exp[-(t/b)^2],$$

$$h(t) = \int_R \exp[-(x/a)^2] \exp[-((t-x)/b)^2] \, dx.$$

Using equation (6.10), we know that

$$\mathcal{F}[\exp[-(t/a)^2]] = (a^2\pi)^{1/2} \exp[-(\pi aw)^2]$$

and

$$\mathcal{F}[\exp[-(t/b)^2]] = (b^2\pi)^{1/2} \exp[-(\pi bw)^2].$$

Thus,

$$H(w) = \mathcal{F}[h(t)] = ab\pi \exp[-(\pi w)^2(a^2 + b^2)],$$

which is obviously another Gaussian function. Now using equation (6.9) and (6.10), we determine the inverse transform of $H(w)$ to be

$$h(t) = \frac{ab(\pi)^{1/2}}{(a^2 + b^2)^{1/2}} \exp\left(\frac{t^2}{a^2 + b^2}\right).$$

Thus, we see that the convolution of two Gaussian functions is also a Gaussian function with $1/e$ radius $r = (a^2 + b^2)^{1/2}$. In other words, the individual $1/e$ radii a and b are combined to yield the new radius $(a^2 + b^2)^{1/2}$.

The analog, or dual, of the convolution theorem is the product theorem which tells us that the Fourier transform of the product of two functions is the convolution of the individual transforms.

Theorem 6.18 (Product). If $|F(w)| \in L(R)$ and $|g(t)| \in L(R)$, where $f(t)$ and $F(w)$ as well as $g(t)$ and $G(w)$ are Fourier transform pairs, then

$$\mathcal{F}[f(t)g(t)] = F(w)*G(w).$$

Proof. Let $h(t) = f(t)g(t)$, from which it follows that

$$H(w) = \mathcal{F}[f(t)g(t)] = \int_R f(t)g(t)e^{-2\pi iwt}\, dt.$$

For the moment we assume that the above integral exists. First, however, we use equation (6.2) and express $f(t)$ as an inverse Fourier transform in the above equation, that is,

$$H(w) = \int_R \left[\int_R F(u)e^{2\pi iut}\, du\right]g(t)e^{-2\pi iwt}\, dt.$$

Now, since $|F(w)| \in L(R)$ and $|g(t)| \in L(R)$ we can use Theorem 3.15 to establish the existence of the previous integral and to also interchange the order of integration in the above integral to obtain

$$H(w) = \int_R F(u)\left[\int_R g(t)e^{2\pi iut}e^{-2\pi iwt}\, dt\right] du.$$

We recognize the inner integral as the Fourier transform of $g(t)\exp[2\pi iut]$, and by the second shifting theorem we obtain $G(w - u)$. When this is substituted back into the above equation we obtain our desired results, that is,

$$H(w) = \int F(u)G(w - u)\, du = F(w)*G(w). \qquad \text{Q.E.D.}$$

CORRELATION

In the previous section we considered the convolution product of two function which turned out to be a law of composition that combined two functions to yield a third. In this section we consider another slightly different composition of two functions, known as *cross-correlation*, which plays a very important role in the theory of stochastic analysis. The *cross-correlation* of two functions $f(t)$ and $g(t)$, denoted as $f(t) \star g(t)$, is mathematically defined as follows:

$$h(t) = f(t) \star g(t) = \int_R f(x)g^*(t+x)dx, \qquad (6.29)$$

where $g^*(t)$ denotes the complex conjugate of the function $g(t)$. Obviously, the functions in the above equation are allowed to be complex. When both functions (in particular, g) are real, then $g^*(t) = g(t)$ and the cross-correlation becomes

$$h(t) = f(t) \star g(t) = \int_R f(x)g(t+x)dx \qquad \text{(real functions)}. \qquad (6.30)$$

Equation (6.29) represents the general definition of cross-correlation. However, in much of our work we are only concerned with real functions, and consequently we then work with equation (6.30).

Recall that when we formed the convolution of two functions, we first rotated the second function and then displaced it by an amount t. However, as can be seen from equation (6.30), when we form the cross-correlation of two real functions we simply displace the second function by an amount t.

Similar to the convolution product, the cross-correlation product is distributive with respect to addition. That is to say,

$$f(t) \star [g(t) + h(t)] = f(t) \star g(t) + f(t) \star h(t)$$

and

$$[f(t) + g(t)] \star h(t) = f(t) \star h(t) + g(t) \star h(t).$$

However, it is not necessarily associative or commutative. To demonstrate noncommutativity of the cross-correlation (of real functions), we proceed as follows:

$$f(t) \star g(t) = \int_R f(x)g(x+t)dx.$$

If we first make the change of variable $y = x + t$, which implies that $x = y - t$ and $dx = dy$, then the above equation becomes

$$f(t) \star g(t) = \int_R f(y-t)g(y)dy = \int_R g(y)f(y-t)dy. \qquad (6.31)$$

Now let us consider the cross-correlation of $g(-t)$ and $f(-t)$, that is,

$$g(-t) \star f(-t) = \int_R g(y)f(y-t)dy. \qquad (6.32)$$

Comparing equations (6.31) and (6.32) we see that

$$f(t) \star g(t) = g(-t) \star f(-t). \qquad (6.33)$$

It is also possible to show that for real functions we have

$$[f(t) \star g(t)] \star h(t) = f(t) \star [g(t) \star h(-t)] \tag{6.34}$$

and

$$f(t) \star [g(t) \star h(t)] = [f(t) \star g(t)] \star h(-t), \tag{6.35}$$

which upholds our statement that the cross-correlation product is not associative.

When a function $f(t)$ is cross-correlatied with itself, the result is known as the *autocorrelation product* and the resulting function is known as the *autocorrelation function*. Mathematically, the autocorrelation of $f(t)$ with itself is given as

$$f(t) \star f(t) = \int_R f(x) f^*(t + x) \, dx; \tag{6.36}$$

when f is a real function, the autocorrelation of $f(t)$ is given as

$$f(t) \star f(t) = \int_R f(x) f(t + x) \, dx \qquad \text{(real function).} \tag{6.37}$$

Just as with convolution, we are interested in the Fourier transform of the cross-correlation product of two functions. Before we begin such a discussion, we first briefly digress and present the following.

Theorem 6.19. If $f(t)$ is an absolutely integrable function with a Fourier transform given by $F(w)$, then the Fourier transform of its complex conjugate function $f^*(t)$ is given as $F^*(-w)$.

Proof. First, since $|f(t)| = |f^*(t)|$ we know that the complex conjugate function $f^*(t)$ is also absolutely integrable and that its Fourier transform exists. We now proceed with the definition of the Fourier transform of $f(t)$, that is,

$$F(w) = \int_R f(t) e^{-2\pi i w t} \, dt.$$

We next take the complex conjugate of both sides of the above equation to obtain

$$F^*(w) = \int_R f^*(t) e^{+2\pi i w t} \, dt.$$

Finally, substitution of $-w$ for w in the above equation yields our desired results, that is,

$$F^*(-w) = \int_R f^*(t) e^{-2\pi i w t} \, dt = \mathcal{F}[f^*(t)]. \qquad \text{Q.E.D.}$$

Now let us consider the Fourier transform of the cross-correlation product.

Theorem 6.20 (Cross-Correlation). If the functions $f(t)$, $g(t)$, and

$h(t) = f(t) \star g(t)$ are absolutely integrable, then $\mathcal{F}[f(t) \star g(t)] = F(w)G^*(w)$, where $F(w)$ and $G(w)$ are the Fourier transforms of $f(t)$ and $g(t)$, respectively.

Proof. By definition, we have

$$h(t) = \int_R f(x)g^*(t+x)\,dx,$$

and therefore

$$\mathcal{F}[h(t)] = \int_R \left[\int_R f(x)g^*(t+x)\,dx \right] e^{-2\pi i w t}\,dt.$$

Since f and g are both absolutely integrable and, consequently, are Lebesgue integrable, we can interchange the order of integration to obtain

$$\int_R f(x) \left[\int_R g^*(t+x)e^{-2\pi i w t}\,dt \right] dx.$$

We note that the inner integral is the Fourier transform of the function $g^*(t+x)$, and by the first shifting theorem we know that it is equal to $G^*(-w)e^{2\pi i w x}$. When this is used in the above equation, we find

$$\mathcal{F}[h(t)] = \int_R f(x)G^*(-w)e^{2\pi i w x}\,dx.$$

Now substituting $-w$ for w in the above equation, we obtain our desired results, that is,

$$\mathcal{F}[h(t)] = \int_R f(x)e^{-2\pi i w x}\,dx\,G^*(w) = F(w)G^*(w). \qquad \text{Q.E.D.}$$

As an immediate consequence of this theorem, we obtain the Fourier transform of the autocorrelation of the function $f(t)$ with itself. This is given in the following.

Corollary 6.20. If $f(t)$ and $f(t) \star f(t)$ are absolutely integrable functions, then $\mathcal{F}[f(t) \star f(t)] = F(w)F^*(w) = \|F(w)\|^2$.

We next consider a theorem that describes the symmetry of the Fourier transform of the cross-correlation product.

Theorem 6.21 (Symmetry). If the Fourier transform of the cross-correlation function exists, then it is an even function, that is, $H(-w) = H(w)$.

Proof. By definition, we have

$$h(t) = \int_R f(x)g^*(t+x)\,dx,$$

from which we obtain

$$H(w) = \int_R \left[\int_R f(x)g^*(t+x)\,dx \right] e^{-2\pi iwt}\,dt.$$

We now let $t + x = y$, which implies that $t = y - x$ and $dt = dy$; thus,

$$H(w) = \int_R \int_R f(x)g^*(y)e^{-2\pi iw(y-x)}\,dy\,dx,$$

$$H(w) = \int_R f(x)e^{2\pi iwx}\,dx \int_R g^*(y)e^{-2\pi iwy}\,dy,$$

$$H(w) = F(-w)G^*(-w).$$

However, from Theorem 6.20 we know that

$$H(w) = F(w)G^*(w)$$

or

$$H(-w) = F(-w)G^*(-w) = H(w). \qquad \text{Q.E.D.}$$

Using the results of Corollary 6.20, we are able to demonstrate *Parseval's energy formula* for Fourier transforms. From this corollary we know

$$\mathcal{F}^{-1}[\|F(w)\|^2] = f(t) \star f(t) = \int_R f(x)f^*(t+x)\,dx$$

or

$$\int_R \|F(w)\|^2 e^{2\pi iwt}\,dx = \int_R f(x)f^*(t+x)\,dx.$$

Finally, if we let $t = 0$ in the above equation, we obtain

$$\int_R \|F(w)\|^2\,dw = \int_R \|f(t)\|^2\,dt. \qquad (6.36)$$

This result often enables us to simplify the calculation of certain infinite integrals, as illustrated in the following.

Example 18. Let us use Parseval's energy formula (equation (6.36)) to determine the value of the following infinite integral:

$$\int_{-\infty}^{\infty} \text{sinc}^2(2\pi wa)\,dw.$$

We first recall that $p_a(t)$ and $\text{sinc}(2\pi wat)$ are Fourier transform pairs. Therefore,

$$\int_{-\infty}^{\infty} \text{sinc}^2(2\pi wa)\,dw = \int_{-\infty}^{\infty} p_a(t)\,dt = \int_{-a}^{a} dt = 2a.$$

SELF-RECIPROCITY AND THE HERMITE FUNCTIONS

Earlier in this chapter (Example 3) we introduced the Gaussian function. One of the unique properties of this Gaussian function was the fact that its Fourier transform was also a Gaussian function. This property is known as *self-reciprocity*. As it turns out, the Gaussian function is a special case of a much larger class of functions known as the *Hermite functions*. These functions are formed as the product of the Hermite polynomials $H_n(t)$ and the simple Gaussian term $\exp(-\pi t^2)$. The Hermite polynomials are defined by the following generating equation:

$$H_n(t) = \left(\frac{-1}{(4\pi)^{1/2}}\right)^n \exp(2\pi t^2)\frac{d^n}{dt^n}\exp(-2\pi t^2). \tag{6.38}$$

Using this generating equation, we determine the first six Hermite polynomials to be given as

$$H_0(t) = 1, \tag{6.39}$$

$$H_1(t) = (4\pi)^{1/2}t, \tag{6.40}$$

$$H_2(t) = 4\pi t^2 - 1, \tag{6.41}$$

$$H_3(t) = [(4\pi)^{1/2}t]^3 - (4\pi)^{1/2}t, \tag{6.42}$$

$$H_4(t) = (4\pi)^2 - 6(4\pi)t^2 + 3, \tag{6.43}$$

$$H_5(t) = [(4\pi)^{1/2}t]^5 - 10[(4\pi)^{1/2}t]^3 + 15(4\pi)^{1/2}t. \tag{6.44}$$

The first six Hermite functions (i.e., $H_n(t)\exp(-\pi t^2)$, $n = 0,\ldots,5$) are shown in Figure 6.19. From this figure and equations (6.39)–(6.44) we note the obvious fact that $H_{2n}(t)\exp(-\pi t^2)$ is an even function, whereas $H_{2n+1}(t)\exp(-\pi t^2)$ is an odd function. It can also be shown that the roots of $H_n(t)$ are always real and simple.

The following recursive formulas for the Hermite polynomials are very useful:

$$H_{n+1}(t) = (4\pi)^{1/2}tH_n(t) - nH_{n-1}(t), \tag{6.45}$$

$$\frac{dH_n(t)}{dt} = n(4\pi)^{1/2}H_{n-1}(t). \tag{6.46}$$

We have already demonstrated (Example 3) that

$$\mathcal{F}[H_0(t)\exp(-\pi t^2)] = \mathcal{F}[\exp(-\pi t^2)] = \exp(-\pi w^2).$$

Now, let us consider the case for $n = 1$, that is,

$$\mathcal{F}[H_1(t)\exp(-\pi t^2)] = \mathcal{F}[(4\pi)^{1/2}t\exp(-\pi t^2)] = (4\pi)^{1/2}\mathcal{F}[t\exp(-\pi t^2)].$$

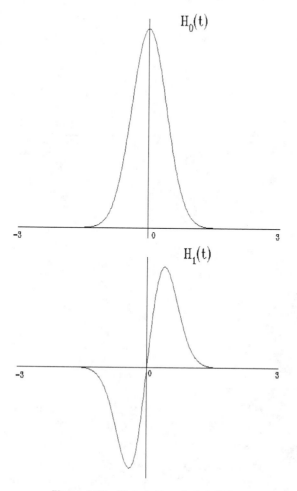

Figure 6.19 First six Hermite functions.

To determine this transform, we use the first derivative theorem (Theorem 6.8) to obtain

$$(4\pi)^{1/2}\mathcal{F}[t\exp(-\pi t^2)] = (4\pi)^{1/2}\frac{i}{2\pi}\frac{d[\exp(-\pi w^2)]}{dw},$$

$$(4\pi)^{1/2}\mathcal{F}[t\exp(-\pi t^2)] = (4\pi)^{1/2}\frac{i}{2\pi}(-2\pi w)\exp(-\pi w^2).$$

Thus,

$$\mathcal{F}[H_1(t)\exp(-\pi t^2)] = -iH_1(w)\exp(-\pi w^2).$$

We now use mathematical induction to demonstrate the self-reciprocity of the Hermite functions in general.

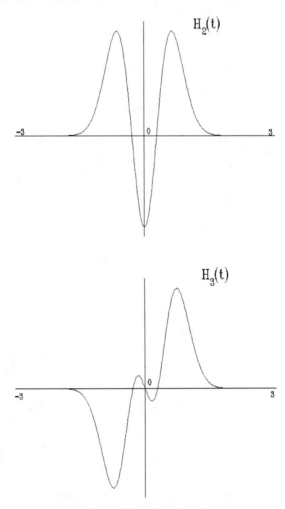

Figure 6.19   *(Continued)*

Theorem 6.22. The Hermite functions exhibit self-reciprocity with respect to the Fourier transform, that is,

$$\mathcal{F}[H_n(t)\exp(-\pi t^2)] = (-i)^n H_n(w)\exp(-\pi w^2).$$

Proof. To apply mathematical induction, we assume that the result is true for $H_n(t)\exp(-\pi t^2)$ and $H_{n-1}(t)\exp(-\pi t^2)$ and show this implies that it must also be true for $H_{n+1}(t)\exp(-\pi t^2)$. We note that since we have already demonstrated the result for $n = 0$ and $n = 1$, this proof establishes the results for all n.

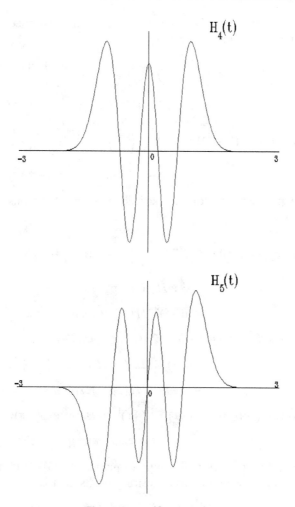

Figure 6.19   *(Continued)*

We begin by using the recrusion formula of equation (6.45) and the linearity property of the Fourier transform to write

$$\mathcal{F}[H_{n+1}(t)\exp(-\pi t^2)] = \mathcal{F}[(4\pi)^{1/2}tH_n(t)\exp(-\pi t^2) - nH_{n-1}(t)\exp(-\pi t^2)],$$
$$\mathcal{F}[H_{n+1}(t)\exp(-\pi t^2)] = (4\pi)^{1/2}\mathcal{F}[tH_n(t)\exp(-\pi t^2)] - n\mathcal{F}[H_{n-1}(t)\exp(-\pi t^2)].$$

Now, by assumption, we have

$$\mathcal{F}[H_{n-1}(t)\exp(-\pi t^2)] = (-i)^{n-1}H_{n-1}(w)\exp(-\pi w^2)$$

and

$$\mathcal{F}[H_n(t)\exp(-\pi t^2)] = (-i)^n H_n(w)\exp(-\pi w^2).$$

When these facts are used along with the first derivative theorem (Theorem 6.8) in the above equation we obtain

$$\mathcal{F}[H_{n+1}(t)\exp(-\pi t^2)] = (4\pi)^{1/2}\frac{i}{2\pi}\frac{d[(-i)^n H_n(w)\exp(-\pi w^2)]}{dw}$$

$$- n(-i)^{n-1}H_{n-1}(w)\exp(-\pi w^2).$$

$$\mathcal{F}[H_{n+1}(t)\exp(-\pi t^2)] = (-i)^{n-1}\left(\frac{1}{\pi}\right)^{1/2}\frac{d[H_n(w)\exp(-\pi w^2)]}{dw}$$

$$- n(-i)^{n-1}H_{n-1}(w)\exp(-\pi w^2).$$

Using the product rule for the derivative of two functions and equation (6.46), we obtain

$$\mathcal{F}[H_{n+1}(t)\exp(-\pi t^2)] = (-i)^{n-1}\left(\frac{1}{\pi}\right)^{1/2}[n(4\pi)^{1/2}H_{n-1}(w)\exp(-\pi w^2)$$

$$- 2\pi w H_n(w)\exp(-\pi w^2)]$$

$$- n(-i)^{n-1}H_{n-1}(w)\exp(-\pi w^2),$$

$$\mathcal{F}[H_{n+1}(t)\exp(-\pi t^2)] = (-i)^{n-1}2nH_{n-1}(w)\exp(-\pi w^2)$$

$$- (-i)^{n-1}2\pi^{1/2}w H_n(w)\exp(-\pi w^2)$$

$$- n(-i)^{n-1}H_{n-1}(w)\exp(-\pi w^2),$$

$$\mathcal{F}[H_{n+1}(t)\exp(-\pi t^2)] = -(-i)^{n-1}[(4\pi)^{1/2}w H_n(w)\exp(-\pi w^2)$$

$$- nH_{n-1}(w)\exp(-\pi w^2)].$$

Now, using the recursion formula (equation (6.45)) and the fact that $-(-i)^{n-1} = (-i)^{n+1}$, we obtain our desired results, that is,

$$\mathcal{F}[H_{n+1}(t)\exp(-\pi t^2)] = (-i)^{n+1}H_{n+1}(w)\exp(-\pi w^2). \qquad \text{Q.E.D.}$$

SUMMARY

In this chapter we discussed the Fourier transform of a function. Specifically, we limited our attention to a class of rather well-behaved functions known as *absolutely integrable functions*. For these functions we were able to establish existence, boundedness, and continuity of the transform. We were also able to describe the behavior of the transform as the frequency variable approached infinity. Properties of the Fourier transform, such as variable shifts, derivatives, and scale change, were examined. This examination clearly pointed to a duality between the function and its transform. In addition, the study of these properties illuminated how they could be used to our advantage to determine the Fourier transform of various functions without having to perform the integral evaluation of equation (6.1). The uniqueness of the Fourier transform was also discussed, and it was learned that only conditions of continuity placed on the function could guarantee that it could be uniquely returned from its Fourier transform.

The convolution product of two functions was discussed both mathematically and graphically. From a physical (or applications) point of view, convolution and the convolution theorem (Theorem 6.17) are perhaps the most important concepts in Fourier analysis. The cross-correlation product, which plays an important role in stochastic analysis, was also studied.

Much of the material presented in this chapter was analytical, or theoretical, in nature. We did, however, purposely provide a rather rich assortment of example problems so that these theoretical discussions could be related to the more practical aspects of Fourier analysis.

Inasmuch as this chapter was devoted to the study of Fourier transforms of absolutely integrable functions, many other useful functions were not discussed. Inherent to the understanding of Fourier analysis are the simple sine and cosine functions, which unfortunately do not fall into the class of absolutely integrable functions. Also absent from this class of functions is the very useful impulse function. These will be studied in the next chapter when we deal with the Fourier transform of distributions.

PROBLEMS

1 Determine the Fourier transform of the function $f(t) = \exp(-a|t|)$.

2 Consider the function $f(t) = \cos(\pi t)$ over the domain $[-1/2, 1/2]$ and defined as zero elsewhere. What is the Fourier transform of this function?

3 Consider the function $f(t) = t$ over the domain $[-1, 1]$ and defined as zero elsewhere. What is the Fourier transform of this function?

4 Consider the function $f(t) = t^2$ over the domain $[0, 1]$ and defined as zero elsewhere. What is the Fourier transform of this function?

5 Consider the function $f(t) = 1 + \cos(\pi t)$ over the domain $[-1, 1]$ and defined as zero elsewhere. What is the Fourier transform of this function?

6 Use the integral definition of the Fourier transform (equation (6.1)) to determine the transform of the triangle function of half-width $2a$ as shown in Figure 6.16. Check your answer with the results obtained in Example 15.

7 Prove the following complex identity:

$$1 - e^{-2\pi i h t} = 2ie^{-iht}\sin(\pi h t).$$

8   Using the first shifting theorem (Theorem 6.5), determine the Fourier transform of the shifted one-sided exponential function $f(t) = \exp[-t + a]$.

9   What are the real and imaginary components of the Fourier transform of the Gaussian function $f(t) = \exp[-\pi t^2]$ that has been shifted to the right by an amount a?

10   If $\mathcal{F}[f(t)] = F(w)$, what is the Fourier transform of the function $g(t) = f(t + a) + f(t - a)$?

11   What is the Fourier transform of $f(t) = e^{-at} \sin(2\pi w_0 t)$ defined over the interval $[0, \infty)$?

12   What is the Fourier transform of $f(t) = e^{-at} \sin(2\pi w_0 t + \nu)$ defined over the interval $[0, \infty)$?

13   Use Theorem 6.7 to determine the Fourier transform of the compressed pulse function $p_a(3t)$. Check your answer by performing the integration of equation (6.1).

14   Use the second derivative theorem (Theorem 6.10) to determine the Fourier transform of the derivative of $-e^{-at}$. Now calculate this transform using the integral expression of equation (6.1). Do your answers agree? (Why?)

15   What is the derivative of the pulse function $p_a(t)$? Can we determine its Fourier transform? (Why?)

16   Use equations (6.19) and (6.20) to determine the real and imaginary portions of the Fourier transform of $p_a(t)$.

17   Prove parts (b), (c), and (d) of Theorem 6.16.

18   Use the integral definition of equation (6.28) to show that the convolution product is associative; that is, show that

$$f(t) * [g(t) * h(t)] = [f(t) * g(t)] * h(t).$$

19   Use the results of Theorem 6.17 and work in the frequency domain to prove that convolution is associative.

20   What is the Fourier transform of the convolution to two equal triangle functions of half-width $2a$?

21   What is the Fourier transform of the convolution of a pulse function of half-width a and a triangle function of half-width b?

22   Use the integral definition of equation (6.30) to establish the results of equation (6.34) concerning the associativity of the cross-correlation product.

23 Use the results of Theorem 6.20 and work in the frequency domain to establish the results of equation (6.33) concerning the noncommutativity of the cross-correlation product.

24 Use the results of Theorem 6.20 and work in the frequency domain to establish the results of equation (6.34) concerning the nonassociativity of the cross-correlation product.

25 Show that when $f(t)$ is a real function, the maximum value of $f(t) \star f(t)$ occurs at $t = 0$.

26 Use Parseval's energy formula (equation (6.36)) to evaluate the following integral:

$$\int_R \frac{1}{(a^2 + 4\pi^2 w^2)^2}\, dw.$$

27 Start with $H_0(t) = 1$ and $H_1(t) = (4\pi)^{1/2}t$ and use the recursive formula (equation (6.45)) to obtain expressions for $H_2(t)$ through $H_5(t)$. Check your answers against equations (6.41)–(6.44).

28 Use equation (6.46) to obtain the derivative of $H_1(t)$ through $H_4(t)$. Check your answers using standard differentiation techniques.

BIBLIOGRAPHY

Papoulis, A., *Systems and Transforms with Applications in Optics*, McGraw-Hill, New York, 1968.

Titchmarsh, E. C., *Fourier Transforms*, Clarendon Press, Oxford, 1948.

Tolstov, G. P., *Fourier Series*, Prentice-Hall, Englewood Cliffs, N.J., 1962.

Weaver, H. J., *Applications of Discrete and Continuous Fourier Analysis*, John Wiley & Sons, New York, 1983.

Worsnop, B. L., *An Introduction to Fourier Analysis*, John Wiley & Sons, New York, 1961.

7

FOURIER TRANSFORM OF A DISTRIBUTION

In this chapter we consider the Fourier transform of a distribution which was defined in Chapter 4. This allows us to round out, or complete, our study of Fourier transforms to include such functions as the sine, cosine, delta, and comb. In Chapter 6 we presented several theorems that described the properties of the Fourier transform of absolutely integrable functions. In this chapter we generalize those results to include the properties of the Fourier transform of distributions.

Recall from Chapter 4 that we defined a distribution $f(t)$ as a continuous linear mapping of functions $f(t) \in S$ to the set of complex numbers C. This mapping was denoted as

$$\langle f(t), g(t) \rangle = z.$$

When the distribution $f(t)$ was also a function (in the space D), we saw that this above mapping was given as

$$\langle f(t), g(t) \rangle = \int_R f(t) g(t) \, dt.$$

We now define the Fourier transform of a distribution $f(t)$ as the distribution $F(w)$ which maps $G(-w)$ to the same complex number z that $f(t)$ maps $g(t)$. In other words, if $g(t)$ and $G(w)$ are Fourier transform pairs, then the Fourier transform of the distribution $f(t)$ is given as the distribution $F(w)$ such that

$$\langle F(w), G(-w) \rangle = \langle f(t), g(t) \rangle. \tag{7.1}$$

Before we proceed with our discussion of the Fourier transform of a distribution, lets briefly digress and consider properties of $G(w)$. By definition,

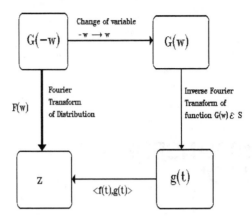

Figure 7.1   Schematic diagram of chain mapping of distribution Fourier transform.

$g(t) \in S$ implies that $g(t)$ possesses all its derivatives and also that $g(t)$ decreases at infinity faster than any power of $1/t$. These facts imply that $|t^n g(t)| \in L(R)$ and also that $|g^{[n]}(t)| \in L(R)$ for all values of n. The derivative theorems (Theorems 6.9 and 6.12) can therefore be applied to show that $G(w) \in S$. Specifically, $|t^n g(t)| \in L(R)$ and Theorem 6.9 imply that $G(w)$ possesses all its derivatives. Also, $|g^{[n]}(t)| \in L(R)$ and Theorem 6.12 imply that $G(w)$ decreases at infinity faster that any power of $1/w$. Thus, we see we have a reciprocity between $g(t)$ and $G(w)$. In other words, if equation (6.1) is used to obtain $G(w)$ from $g(t)$, then $g(t)$ can be uniquely recovered from $G(w)$ by using the inverse Fourier transform formula of equation (6.2).

The fact that both $g(t)$ and $G(w)$ are absolutely integrable functions (as well as functions in the space S) guarantees that we can always find the Fourier transform of a distribution. In other words, it is always possible to find a mapping that will take $G(-w)$ to the complex number $z = \langle f(t), g(t) \rangle$. This is actually accomplished by a chain mapping. First we map $G(-w)$ to $G(w)$ by a simple change of variable ($-w$ for w) mapping. Next we use the inverse Fourier transform to map $G(w)$ to $g(t)$, and finally we use the original distribution $f(t)$ to map $g(t)$ to z. This chain mapping is shown schematically in Figure 7.1. It is a relatively simple matter to demonstrate that this chain mapping is both linear and continuous and that is consequently satisfies the definition of a distribution. It should also be rather obvious that we can always use this chain mapping in reverse to obtain $F(w)$ from $f(t)$. Consequently, $f(t)$ can be considered the inverse Fourier transform of $F(w)$. We use the same notation as in Chapter 6 to denote that the distributions $f(t)$ and $F(w)$ are Fourier transform pairs, that is,

$$F(w) = \mathcal{F}[f(t)] \quad \text{and} \quad f(t) = \mathcal{F}^{-1}[F(w)].$$

As our first result, let us determine the Fourier transform of the delta distribution. By definition, this distribution is given as

$$\langle \delta(t), g(t) \rangle = g(0). \tag{7.2}$$

If we denote the Fourier transform of the delta distribution as $P(w)$, then by equation (7.1) we know

$$\langle P(w), G(-w) \rangle = g(0).$$

Now let us use the inverse Fourier transform expression of equation (6.2) to write

$$g(t) = \int_R G(w) e^{2\pi i w t} \, dw.$$

or

$$g(0) = \int_R G(w) \, dw.$$

If we now substitute $-w$ for w and make the necessary change of signs and limits, we obtain

$$g(0) = \int_R G(-w) \, dw = \int_R 1 G(-w) \, dw = \langle 1, G(-w) \rangle. \tag{7.3}$$

Combining equations (7.2) and (7.3), we find

$$\langle \delta(t), f(t) \rangle = g(0) = \langle 1, G(-w) \rangle.$$

Therefore, we conclude that

$$\mathcal{F}[\delta(t)] = 1. \tag{7.4}$$

This is the same result that we obtained in Chapter 1 when we approached the problem as the limit of a sequence of functions. Because of the reciprocity of the Fourier transforms of distributions, we also obtain

$$\mathcal{F}^{-1}[1] = \delta(t).$$

Let us now consider the situation when the distribution $f(t)$ is also a Lebesgue integrable function. In this case we write

$$\langle f(t), g(t) \rangle = \int_R f(t) g(t) \, dt = \langle F(w), G(-w) \rangle = \int_R F(w) G(-w) \, dw.$$

If we now use the integral expression for the inverse Fourier transform of $g(t)$ in the first integral in above equation, we find

$$\langle F(w), G(w) \rangle = \int_R f(t) \int_R G(w) e^{2\pi i w t} \, dw \, dt.$$

Since both $f(t)$ and $G(w)$ are Lebesgue integrable functions, we can interchange the order of integration to obtain

$$\int_R \int_R f(t) e^{2\pi i w t} \, dt \, G(w) \, dw.$$

Now, substitution of $-w$ for w in the above equation results in

$$\int_R F(w)G(-w)\,dw = \int_R \int_R f(t)e^{-2\pi iwt}\,dt\,G(-w)\,dw.$$

Thus we see that

$$F(w) = \int_R f(t)e^{-2\pi iwt}\,dt,$$

which agrees with the definition presented in the previous chapter. Thus we may consider the Fourier transform of a Lebesgue function to be a special case of the Fourier transform of a distribution.

LINEARITY AND SCALE CHANGE

In this section we consider the properties of linearity and scale change of the Fourier transform of a distribution. We begin with linearity.

Theorem 7.1 (Linearity). If the distributions $F_1(w)$ and $F_2(w)$ are the Fourier transforms of the distributions $f_1(t)$ and $f_2(t)$, respectively, and a and b are any two constants (perhaps complex), then the Fourier transform of the distribution $af_1(t) + bf_2(t)$ is given as $aF_1(w) + bF_2(w)$.

Proof. By assumption, we have

$$\langle f_1(t),g(t)\rangle = \langle F_1(w),G(-w)\rangle$$

and

$$\langle f_2(t),g(t)\rangle = \langle F_2(w),G(-w)\rangle.$$

Inasmuch as distributions are linear forms (see equations (4.6) and (4.7)), we have

$$\langle af_1(t) + bf_2(t),g(t)\rangle = a\langle f_1(t),g(t)\rangle + b\langle f_2(t),g(t)\rangle,$$

$$\langle af_1(t) + bf_2(t),g(t)\rangle = a\langle F_1(w),G(-w)\rangle + b\langle F_2(w),G(-w)\rangle,$$

$$\langle af_1(t) + bf_2(t),g(t)\rangle = \langle aF_1(w) + bF_2(w),G(-w)\rangle,$$

Thus, by definition, we arrive at

$$\mathcal{F}[af_1(t) + bf_2(t)] = aF_1(w) + bF_2(w). \qquad \text{Q.E.D.}$$

We note that this is the same type of result obtained in Chapter 6 when we dealt with absolutely integrable functions.

Before we proceed with the next theorem concerning the Fourier transform of a distribution and scale change, let us recall a basic result from Chapter 4. Given any distribution $h(t)$, the following two relations are true:

$$\langle h(at),g(t)\rangle = \frac{1}{|a|}\langle h(t),g(t/a)\rangle, \qquad (7.5)$$

or, using $1/a$ for a in equation (7.5), we find

$$\langle h(t/a), g(t) \rangle = |a| \langle h(t), g(at) \rangle. \tag{7.6}$$

We now prove the following.

Theorem 7.2 (Scale Change). If the distributions $f(t)$ and $F(w)$ are Fourier transform pairs, then

$$\mathcal{F}[f(at)] = \frac{F(w/a)}{|a|}.$$

Proof. Using equation (7.5), we have

$$\langle f(at), g(t) \rangle = 1/|a| \langle f(t), g(t/a) \rangle.$$

We first recall (Theorem 6.7) that the Fourier transform of the function $g(t/a)$ is given as $|a|G(wa)$. Therefore, taking the Fourier transform of both sides of the above equation yields

$$\langle \mathcal{F}[f(at)], G(-w) \rangle = \langle F(w), G(-wa) \rangle.$$

Now, using equation (7.6) on the right-hand side of the above equation, we find

$$\langle \mathcal{F}[f(at)], G(-w) \rangle = (1/|a|) \langle F(w/a), G(-w) \rangle = \langle F(w/a)/|a|, G(-w) \rangle.$$

Thus we obtain our desired result, that is,

$$\mathcal{F}[f(at)] = F(w/a)/|a|. \qquad\qquad \text{Q.E.D.}$$

An immediate consequence of this theorem is obtained by setting $a = -1$, in which case we obtain the following.

Corollary 7.2. If the distributions $f(t)$ and $F(w)$ are Fourier transform pairs, then we have $\mathcal{F}[f(-t)] = F(-w)$.

Again we note that this is the same result obtained in Chapter 6 when we dealt with the scale change of a function that was absolutely integrable.

THE SHIFTING THEOREMS

In this section we show that the first and second shifting theorems which were demonstrated in Chapter 6 for absolutely integrable functions are also valid for distributions.

Theorem 7.3 (First Shifting Theorem). If the distributions $f(t)$ and $F(w)$ are Fourier transform pairs, then the Fourier transform of $f(t - a)$ is given by the distribution $F(w) \exp[-2\pi i w a]$.

Proof. By definition (equation (4.9)), we know that

$$\langle f(t-a), g(t) \rangle = \langle f(t), g(t+a) \rangle$$

We first note that the Fourier transform of the function $g(t+a)$ is given as $G(w)\exp[2\pi iwa]$ (Theorem 6.5). Now, taking the Fourier transform of both sides of the above equation, we obtain

$$\langle \mathcal{F}[f(t-a)], G(-w) \rangle = \langle F(w), G(-w)e^{-2\pi iwa} \rangle.$$

However, by equation (4.8) (product of a function and a distribution), we have

$$\langle F(w), G(-w)e^{-2\pi iwa} \rangle = \langle F(w)e^{-2\pi iwa}, G(-w) \rangle,$$

or

$$\langle \mathcal{F}[f(t-a)], G(-w) \rangle = \langle F(w)e^{-2\pi iwa}, G(-w) \rangle.$$

Thus we obtain our desired results, that is,

$$\mathcal{F}[f(t-a)] = F(w)e^{-2\pi iwa}. \qquad \text{Q.E.D.}$$

We now present the dual of the first shifting theorem (Theorem 7.3).

Theorem 7.4 (Second Shifting Theorem). If the distributions $f(t)$ and $F(w)$ are Fourier transform pairs, then the Fourier transform of $f(t)\exp[2\pi iat]$ is given by $F(w-a)$.

Proof. Using equation (4.8), which describes the product of a function and a distribution, we have

$$\langle f(t)\exp[2\pi iat], g(t) \rangle = \langle f(t), g(t)\exp[2\pi iat] \rangle.$$

By Theorem 6.6 we know that the Fourier transform of the function $g(t)\exp[2\pi iat]$ is given as $G(w-a)$. Therefore, taking the Fourier transform of both sides of the above equation yields

$$\langle \mathcal{F}[f(t)\exp[2\pi at], g(t) \rangle = \langle F(w), G(-w+a) \rangle = \langle F(w-a), G(-w) \rangle.$$

Thus,

$$\mathcal{F}[f(t)\exp[2\pi at]] = F(w-a). \qquad \text{Q.E.D.}$$

As a direct consequence of Theorem 7.4, we have the following corollary.

Corollary 7.4. If the distributions $f(t)$ and $F(w)$ are Fourier transform pairs, then we have

(a) $\mathcal{F}[f(t)\cos(2\pi at)] = [F(w+a) + F(w-a)]/2$, and
(b) $\mathcal{F}[f(t)\sin(2\pi at)] = i[F(w+a) - F(w-a)]/2$.

Proof. Using Theorem 7.4, we know

$$\mathcal{F}[f(t)\cos(2\pi at) + if(t)\sin(2\pi at)] = F(w - a),$$

$$\mathcal{F}[f(t)\cos(2\pi at) - if(t)\sin(2\pi at)] = F(w + a).$$

Addition of the two previous distributions yields the results of part (a), whereas subtraction yields the results of part (b).   Q.E.D.

Example 1. We have already demonstrated that the Fourier transform of the delta distribution $\delta(t)$ is the unit constant function $F(w) = 1$. Applying Theorem 7.2 we determine the Fourier transform of the shifted delta function as follows:

$$\mathcal{F}[\delta(t - a)] = F(w)e^{-2\pi iaw} = 1e^{-2\pi iaw}.$$

Thus,

$$\mathcal{F}[\delta(t - a)] = e^{-2\pi iaw}. \tag{7.7}$$

We now derive a very useful theorem concerning the Fourier transform of a Fourier transform of a distribution.

Theorem 7.5 (Transform of a Transform). If $f(t)$ is a distribution, then $\mathcal{F}[\mathcal{F}[f(t)]] = f(-t)$.

Proof. From Chapter 4 we know

$$\langle f(-t), g(t) \rangle = \langle f(t), g(-t) \rangle.$$

Taking the Fourier transform of the right-hand side of the above equation and noting the fact that $\mathcal{F}[g(-t)] = G(-w)$, we obtain

$$\langle f(-t), g(t) \rangle = \langle \mathcal{F}[f(t)], G(w) \rangle.$$

Now, once again taking the Fourier transform of the right-hand side of the above equation and using Theorem 6.12 (i.e., $\mathcal{F}[G(-w)] = g(-t)$), we obtain

$$\langle f(-t), g(t) \rangle = \langle \mathcal{F}[\mathcal{F}[f(t)]], g(t) \rangle.$$

Thus we obtain our desired results, that is,

$$\mathcal{F}[f(t)] = f(-t). \qquad\text{Q.E.D.}$$

Example 2. We now demonstrate the fact that the Fourier transform of the unit constant function (distribution) is the delta distribution. This is easily accomplished by application of Theorem 7.5. In other words, we know that

$$\mathcal{F}[\delta(t)] = 1.$$

Now, taking the Fourier transform of the above equation and applying Theorem 7.5, we find

$$\mathcal{F}[\mathcal{F}[\delta(t)]] = \delta(-t) = \delta(t) = \mathcal{F}[1].$$

Thus we see

$$\mathcal{F}[1] = \delta(t). \tag{7.8}$$

Example 3. In this example we consider the Fourier transforms of the sine and cosine functions. From Corollary 7.4(a) we have

$$\mathcal{F}[f(t)\cos(2\pi at)] = [F(w+a) + F(w-a)]/2.$$

If we now let $f(t) = 1$ (which implies $F(w) = \delta(w)$) in the above equation, we obtain

$$\mathcal{F}[\cos(2\pi at)] = [\delta(w+a) + \delta(w-a)]/2. \tag{7.9}$$

Similar logic yields

$$\mathcal{F}[\sin(2\pi at)] = i[\delta(w+a) - \delta(w-a)]/2. \tag{7.10}$$

These transforms are shown in Figure 7.2 ((a) cosine and (b) sine). We note that the frequency content of each transform is a single frequency term located at $+a$ and $-a$. The transform of the cosine is real, whereas that of the sine is imaginary.

Example 4. We now use Theorem 7.5 to determine the Fourier transform of the sinc function. We begin by considering the pulse function $p_a(u)$ in a distribution sense and note that

$$2a\,\text{sinc}(2\pi ax) = \mathcal{F}[p_a(u)].$$

Taking the Fourier transform of both sides of the above equation, we obtain

$$2a\mathcal{F}[\text{sinc}(2\pi ax)] = \mathcal{F}[\mathcal{F}[p_a(u)]] = p_a(-u) = p_a(u).$$

Thus,

$$\mathcal{F}[\text{sinc}(2\pi ax)] = \frac{p_a(u)}{2a}.$$

Now, using the scale change theorem (Theorem 7.2) with scale factor $1/2\pi a$, we find

$$\mathcal{F}[\text{sinc}(x)] = \pi p_a(2\pi au).$$

We note that $p_a(2\pi au)$ is equal to unity whenever $2\pi au \leq a$ or $u \leq 1/2\pi$ (and equal to zero elsewhere). Using this fact and changing the variables (for notational consistency) x to t and $2\pi au$ to w, we obtain

$$\mathcal{F}[\text{sinc}(t)] = \pi p_{1/2\pi}(w). \tag{7.11}$$

THE DERIVATIVE THEOREMS

In Chapter 4 we defined the derivative of a distribution as that distribution which mapped $g(t)$ to the negative value of the complex number to which the original distribution mapped $g'(t)$, that is,

$$\langle f'(t), g(t) \rangle = -\langle f(t), g'(t) \rangle.$$

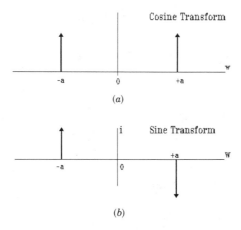

Figure 7.2 Fourier transforms of the (a) cosine and the (b) sine function.

Higher-order derivatives are similarly defined as

$$\langle f^{[n]}(t), g(t)\rangle = (-1)^n \langle f(t), g^{[n]}(t)\rangle.$$

In this section we consider both the Fourier transform of the derivative of a distribution and the derivative of the Fourier transform of a distribution. We begin with the following.

Theorem 7.6 (Derivative of the Transform). If the distributions $f(t)$ and $F(w)$ are Fourier transform pairs, then

$$\frac{dF(w)}{dw} = \mathcal{F}[-2\pi it f(t)].$$

Proof. By definition of the Fourier transform of a distribution, we have

$$\langle f(t), g(t)\rangle = \langle F(w), G(-w)\rangle.$$

Now, using the definition of the derivative, we have

$$\langle F'(w), G(-w)\rangle = -\langle F(w), G'(-w)\rangle.$$

However, using the derivative theorem (Theorem 6.8) for a function, we have

$$\mathcal{F}^{-1}[G'(w)] = -2\pi it g(t), \quad \text{or} \quad \mathcal{F}^{-1}[G'(-w)] = 2\pi it g(t).$$

When this is used in the above equation, we find

$$\langle F'(w), G(-w)\rangle = -\langle f(t), 2\pi it g(t)\rangle = \langle -2\pi it f(t), g(t)\rangle.$$

Thus, by the definition of the Fourier transform, we see that

$$\frac{dF(w)}{dw} = -2\pi i \mathcal{F}[t f(t)]. \qquad\qquad \text{Q.E.D.}$$

This result is easily generalized to read as follows.

Theorem 7.7. If the distributions $f(t)$ and $F(w)$ are Fourier transform pairs, then

$$F^{[n]}(w) = \mathcal{F}[(-2\pi i t)^n f(t)].$$

We now consider the dual of Theorem 7.6, which deals with the Fourier transform of the derivative of a distribution.

Theorem 7.8 (Transform of the Derivative). If the distributions $f(t)$ and $F(w)$ are Fourier transform pairs, then we have

$$\mathcal{F}[f'(t)] = 2\pi i w F(w).$$

Proof. By definition, we have

$$\langle f'(t), g(t) \rangle = -\langle f(t), g'(t) \rangle.$$

Now, taking the Fourier transform of both sides of the above equation and using the fact that $\mathcal{F}[g'(t)] = 2\pi i w G(w)$ (Theorem 6.10), we see that

$$\langle \mathcal{F}[f'(t)], G(-w) \rangle = -\langle F(w), -2\pi i w G(-w) \rangle = \langle 2\pi i w F(w), G(-w) \rangle.$$

Therefore,

$$\mathcal{F}[f'(t)] = 2\pi i w F(w). \qquad\qquad \text{Q.E.D.}$$

This theorem is generalized to read as follows.

Theorem 7.9. If the distributions $f(t)$ and $F(w)$ are Fourier transform pairs, then we have

$$\mathcal{F}[f^{[n]}(t)] = (2\pi i w)^n F(w).$$

We now present several examples illustrating the application of these theorems.

Example 5. We have previously shown (Chapter 6, Example 2) that the Fourier transform of the one-sided exponential function is given as

$$\mathcal{F}[e^{-at}] = \frac{1}{a + 2\pi i w}.$$

The derivative of this function is

$$f'(t) = \delta(t) - ae^{-at}.$$

We note the presence of the delta distribution located at zero is due to the jump discontinuity of the one-sided exponential at zero. By conventional techniques, we determine

$$\mathcal{F}[f'(t)] = \mathcal{F}[\delta(t)] - \mathcal{F}[ae^{-at}],$$

$$\mathcal{F}[f'(t)] = 1 - \frac{a}{a + 2\pi i w}.$$

or, placing both terms over a common denominator, we find

$$\mathcal{F}[f'(t)] = \frac{2\pi i w}{a + 2\pi i w}.$$

This same result is easily obtained using Theorem 7.8, that is,

$$\mathcal{F}[f'(t)] = 2\pi i w F(w) = \frac{2\pi i w}{a + 2\pi i w}.$$

Example 6. Let us now determine the Fourier transform of the function $f(t) = te^{-at}$. Using Theorem 7.6, we find

$$\mathcal{F}[tf(t)] = \left(-\frac{1}{2\pi i}\right)\frac{dF(w)}{dw}.$$

This fact, coupled with the fact that $\mathcal{F}[e^{-at}] = (a + 2\pi i w)^{-1}$, yields

$$\mathcal{F}[te^{-at}] = \left(-\frac{1}{2\pi i}\right)(a + 2\pi i w)^{-2}(-2\pi i) = \frac{1}{(a + 2\pi i w)^2}.$$

Example 7. The derivative of a pulse function of half-width a is given as two delta functions located at $w = -a$ and $w = +a$, that is,

$$p_a'(t) = \delta(t + a) - \delta(t - a).$$

Thus,

$$\mathcal{F}[p_a'(t)] = e^{+2\pi i w a} - e^{-2\pi i w a} = 2i\sin(2\pi w a).$$

Now, using Theorem 7.8 and the fact that

$$\mathcal{F}[p_a(t)] = \frac{2a\sin(2\pi w a)}{2\pi w a},$$

we obtain

$$\mathcal{F}[p_a'(t)] = 2\pi i w \mathcal{F}[p_a(t)] = 2i\sin(2\pi w a).$$

Example 8. Shown in Figure 7.3 is the unit step function which is denoted as stp(t) and defined mathematically as:

$$\text{stp}(t) = \begin{cases} 1, & t \geq 0, \\ 0, & t < 0. \end{cases} \tag{7.12}$$

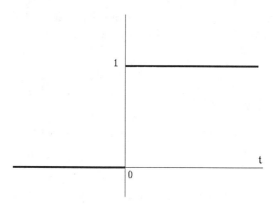

Figure 7.3 The unit step function.

To determine the Fourier transform of this step function let us begin with the Fourier transform of the one-sided exponential function

$$\mathcal{F}[e^{-at}] = \frac{1}{a + 2\pi i w}. \tag{7.13}$$

We note that in the limit as $a \to 0$ we have this one-sided exponential function going to the unit step function, that is,

$$\text{stp}(t) = \lim_{a \to 0} e^{-at}.$$

Therefore, we obtain our desired result by setting $a = 0$ in equation (7.13), that is,

$$\mathcal{F}[\text{stp}(t)] = \frac{1}{2\pi i w}. \tag{7.14}$$

Example 9. Shown in Figure 7.4 is the sgn function which is defined mathematically as:

$$\text{sgn}(t) = \begin{cases} 1, & t \geq 0, \\ -1, & t < 0. \end{cases} \tag{7.15}$$

We are also able to write this function in terms of the previously defined unit step function and unit constant function as follows:

$$\text{sgn}(t) = 2\,\text{stp}(t) - 1.$$

Taking the Fourier transform of both sides of the above equation, we find

$$\mathcal{F}[\text{sgn}(t)] = \frac{1}{i\pi w} - \delta(w). \tag{7.16}$$

Example 10. Using the generalized derivative theorem (Theorem 7.7) and the one-sided exponential function of equation (7.13), we are able to derive the

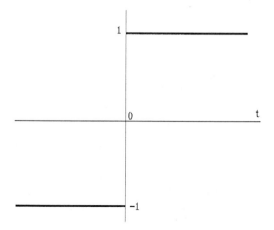

Figure 7.4   The sgn function.

following relationship:

$$\mathcal{F}[t^n e^{-at}] = \frac{n!}{(a + 2\pi i w)^{n+1}}.$$

If we now set $a = 0$ in the above expression, we find

$$\mathcal{F}[t^n] = \frac{n!}{(2\pi i w)^{n+1}}. \qquad (7.17)$$

We note here that the functions $f(t) = t^n$ are considered to be only defined for $t \geq 0$ (and equal to zero for $t < 0$).

SYMMETRY CONSIDERATIONS

In this section we parallel the development presented in Chapter 6 and discuss symmetry properties of certain distributions and their Fourier transforms.

In Chapter 4 we defined what we meant by odd and even distributions. That is to say, a distribution is called even if and only if for every $g(t) \in S$ we have

$$\langle f(t), g(-t) \rangle = \langle f(t), g(t) \rangle.$$

A distribution is called odd if and only if for every $g(t) \in S$ we have

$$\langle f(t), g(-t) \rangle = -\langle f(t), g(t) \rangle.$$

In Chapter 4 we also defined real and imaginary distributions as follows:

A distribution is called real if and only if, for every real function $g(t) \in S$, $\langle f(t), g(t) \rangle$ is a real number.

A distribution is called (pure) imaginary if and only if $\langle f(t),g(t)\rangle$ is an imaginary number.

Finally, we define the complex conjugate of a distribution (denoted as $f^*(t)$) as that distribution that maps $g(t)$ as follows:

$$\langle f^*(t),g(t)\rangle = \langle f(t),g^*(t)\rangle^*.$$

The following two theorems deal with properties of the Fourier transform of odd and even and real and imaginary as well as complex conjugate distributions.

Theorem 7.10. If the distributions $f(t)$ and $F(w)$ are Fourier transform pairs then the Fourier transform of the complex conjugate distribution $f^*(t)$ is given by $F^*(-w)$.

Proof. By definition of the complex conjugate and the Fourier transform of a distribution we have

$$\langle f^*(t),g(-t)\rangle = \langle f(t),g^*(-t)\rangle^*,$$
$$\langle f^*(t),g(-t)\rangle = \langle F(w),G^*(-w)\rangle^* = \langle F(-w),G^*(w)\rangle^*.$$

Now, again using the definition of the complex conjugate, we see that $\langle F^*(-w),G(w)\rangle = \langle F(-w),G^*(w)\rangle^*$. When this is used in the above equation, we obtain our desired results, that is,

$$\langle f^*(t),g(-t)\rangle = \langle F^*(-w),G(w)\rangle. \qquad \text{Q.E.D.}$$

Theorem 7.11 (Symmetry). If the distributions $f(t)$ and $F(w)$ are Fourier transform pairs, then we have:

(a) If $f(t)$ is real and even, then $F(w)$ is real and even.
(b) If $f(t)$ is real and odd, then $F(w)$ is imaginary and odd.
(c) If $f(t)$ is imaginary and even, then $F(w)$ is imaginary and even.
(d) If $f(t)$ is imaginary and odd, then $F(w)$ is real and odd.

Proof. By assumption, $f(t)$ is even and therefore

$$\langle f(t),g(t)\rangle = \langle f(t),g(-t)\rangle.$$

Now, transforming both sides of the above equation, we find

$$\langle F(w),G(-w)\rangle = \langle F(w),G(w)\rangle,$$

which, by definition, implies that $F(w)$ is an even distribution.

Now, since $f(t)$ is real we know that for any real function $g(t)=g^*(t)$ we have

$$\langle f^*(t),g(-t)\rangle = \langle f(t),g^*(-t)\rangle^* = \langle f(t),g(-t)\rangle^*$$

and, by definition, $\langle f(t), g(-t)\rangle^*$ must be a real number and therefore equal to $\langle f(t), g(-t)\rangle$. Thus, the above equation becomes

$$\langle f^*(t), g(-t)\rangle = \langle f(t), g(-t)\rangle,$$

and therefore we see that for a real distribution we have $f^*(t) = f(t)$.

Now, transforming both sides of the above equation, we obtain

$$\langle F^*(-w), G(w)\rangle = \langle F(w), G(w)\rangle.$$

However, since $F(w)$ is an even function we have $F^*(-w) = F^*(w)$. When this is used in the above equation we obtain

$$\langle F^*(w), G(w)\rangle = \langle F(w), G(w)\rangle,$$

and we see that $F^*(w) = F(w)$, which implies that $F(w)$ is a real distribution.

The proofs of parts (b), (c), and (d) are similar and are left as an exercise for the reader.   Q.E.D.

FOURIER TRANSFORM OF THE COMB DISTRIBUTION

In this section we consider the Fourier transform of the comb distribution, which turns out to be another comb distribution. In other words, the comb distribution exhibits the property of self-reciprocity. To properly determine the Fourier transform of a comb distribution, we must first briefly digress and present the following preliminary result.

Consider the function $h(t)$ which is not everywhere zero but, instead, has isolated zeros at the points kT, that is,

$$h(kT) = 0 \qquad \text{for} \quad k = -\infty, \ldots, -1, 0, 1, \ldots, \infty. \tag{7.18}$$

The solution (for $f(t)$) to the equation

$$h(t)f(t) = 0 \tag{7.19}$$

is given as the following distribution:

$$f(t) = \sum_{k=-\infty}^{\infty} A_k \delta(t - kT), \tag{7.20}$$

where the A_k terms are (perhaps complex) constants for each integer value k. To demonstrate this result, we first use the fact that

$$\sum_{k=-\infty}^{\infty} A_k h(kT) g(kT) = 0 = \left\langle \sum_{k=-\infty}^{\infty} A_k \delta(t - kT), h(t)g(t) \right\rangle. \tag{7.21}$$

However, by assumption, we have

$$0 = \langle 0, g(t)\rangle = \langle h(t)f(t), g(t)\rangle = \langle f(t), h(t)g(t)\rangle.$$

Therefore, comparing equations (7.21) and (7.22), we find

$$\langle f(t), h(t)g(t) \rangle = \left\langle \sum_{k=-\infty}^{\infty} A_k \delta(t - kT), h(t)g(t) \right\rangle,$$

or

$$f(t) = \sum_{k=-\infty}^{\infty} A_k \delta(t - kT).$$

We are now in a position to consider the Fourier transform of

$$\mathrm{comb}_T(t) = \sum_{k=-\infty}^{\infty} \delta(t - kT).$$

By definition, we have

$$\left\langle \sum_{k=-\infty}^{\infty} \delta(t - kT), g(t) \right\rangle = \left\langle \mathcal{F}\left[\sum_{k=-\infty}^{\infty} \delta(t - kT) \right], G(-w) \right\rangle.$$

However, using the first shifting theorem (Theorem 7.3), we know

$$\mathcal{F}[\delta(t - kT)] = 1e^{-2\pi ikTw} = e^{-2\pi ikTw}.$$

Thus,

$$\langle \mathrm{comb}_T(t), g(t) \rangle = \left\langle \sum_{k=-\infty}^{\infty} e^{-2\pi ikTw}, G(-w) \right\rangle$$

or

$$\mathcal{F}[\mathrm{comb}_T(t)] = \sum_{k=-\infty}^{\infty} e^{-2\pi ikTw}. \tag{7.23}$$

However, we can write the above summation as

$$\sum_{k=-\infty}^{\infty} e^{-2\pi ikTw} = 1 + 2\cos(2\pi Tw) + 2\cos(2\pi 2Tw) + \cdots,$$

or, referring back to Chapter 5 (equation (5.44)), we see

$$\sum_{k=-\infty}^{\infty} e^{-2\pi ikTw} = 2 \lim_{n \to \infty} D_n(2\pi wT),$$

where D_n is the Dirichlet kernel function.

If we again refer back to Chapter 5 (Figure 5.1), we see that, for all values of n, $D_n(2\pi wT)$ is a periodic function of period $1/T$. Furthermore, from this figure we see that in the limit as $n \to \infty$, $D_n(2\pi wT)$ "seems to" approach a series of equally spaced (spacing $= 1/T$) delta distributions. In other words, $D_n(2\pi wT)$ "seems to" approach a comb distribution with spacing $1/T$. Let us now rigorously demonstrate this fact. For convenience we temporarily denote the Fourier transform of $\text{comb}_T(t)$ as $T(w)$. We first note that $\text{comb}_T(t)$ is periodic with period T, and thus we write

$$\text{comb}_T(t) = \text{comb}_T(t+T)$$

and

$$\langle \text{comb}_T(t), g(t) \rangle = \langle \text{comb}_T(t+T), g(t) \rangle.$$

Taking the Fourier transform of both sides of the above equation and using the first shifting theorem (Theorem 7.3) on the right-hand side, we obtain

$$\langle T(w), G(-w) \rangle = \langle T(w)e^{-2\pi i T w}, G(-w) \rangle$$

or

$$\langle T(w)[1 - e^{-2\pi i T w}], G(-w) \rangle = 0.$$

Therefore, we conclude that

$$T(w)[1 - e^{-2\pi i T w}] = 0.$$

We note that the above equation is of the form given by equation (7.19), in which $1 - \exp[-2\pi i T w]$ has isolated zeros at $w = j/T$ $[j = -\infty, \ldots, -1, 0, 1, \ldots, \infty]$. Therefore, we know that $T(w)$ must be of the form

$$T(w) = \mathcal{F}[\text{comb}_T(t)] = \sum_{j=-\infty}^{\infty} A_j \delta \left(w - \frac{j}{T} \right). \tag{7.24}$$

This equation does indeed show that the transform of the comb distribution is a series of delta functions spaced $1/T$ frequency units apart. We next show that $A_j = 1/T$ for all values of j, and consequently

$$\mathcal{F}[\text{comb}_T(t)] = \frac{1}{T} \sum_{j=-\infty}^{\infty} \delta \left(w - \frac{j}{T} \right) = \frac{1}{T} \text{comb}_{1/T}(w). \tag{7.25}$$

We begin by combining equations (7.23) and (7.24) to obtain

$$\sum_{k=-\infty}^{\infty} e^{-2\pi i k T w} = \sum_{j=-\infty}^{\infty} A_j \delta \left(w - \frac{j}{T} \right),$$

or

$$1 + \sum_{k=-\infty}^{\infty} 2\cos(2\pi k T w) = \sum_{j=-\infty}^{\infty} A_j \delta \left(w - \frac{j}{T} \right). \tag{7.26}$$

Our desired results are obtained by integrating equation (7.26) over the frequency intervals $(j/T - 1/2T, \ j/T + 1/2T)$ for $j = -\infty, \ldots, -1, 0, 1, \ldots, \infty)$, that is,

$$\int_{j/T-1/2T}^{j/T+1/2T} dw + \sum_{k=-\infty}^{\infty} \int_{j/T-1/2T}^{j/T+1/2T} 2\cos(2\pi k T w)\, dw$$

$$= \sum_{j=-\infty}^{\infty} A_j \int_{j/T-1/2T}^{j/T+1/2T} \delta\left(w - \frac{j}{T}\right) dw.$$

The first integral is equal to $1/T$. For all values of j, the second integral is taken over one period $(1/T)$ of each cosine term and is consequently equal to zero. Finally, from Chapter 4 we know that the integral from 0^- to 0^+ of $\delta(w)$ is unity. Thus

$$A_j \int_{j/T-1/2T}^{j/T+1/2T} \delta\left(w - \frac{j}{T}\right) dw = A_j \int_{0-}^{0+} \delta(w)\, dw = A_j,$$

and we see $1/T = A_j$ for all values of j.

CONVOLUTION OF DISTRIBUTIONS

In this section we consider the convolution product of two distributions. As we learned in Chapter 6, for absolutely integrable functions the convolution product of two time domain functions and the simple product of their transfoms in the frequency domain are closely related via the convolution theorem (Theorem 6.17). In this section we demonstrate the same type of result for distributions. In addition we present criteria by which we are able to determine when the convolution of two distributions exists.

We define the convolution of two distributions $f(t)$ and $h(t)$ (denoted as $f(t)*h(t)$) as

$$\langle f(t)*h(t), g(t) \rangle = \langle f(t), h(-t)*g(t) \rangle. \tag{7.27}$$

In words, the convolution product $f(t)*h(t)$ is a distribution that maps $g(t) \in L(R)$ to the same complex number that the distribution $f(t)$ maps $h(-t)*g(t)$. This definition is valid when the above mappings exist. Before we look deeper into the question of existence let us consider equation (7.27) when both $f(t)$ and $h(t)$ are absolutely, or Lebesgue, integrable functions. In this case the above inner products and convolution products are considered in the integral sense, that is,

$$\left\langle \int_R f(x)h(t-x)dx, g(t) \right\rangle = \left\langle f(t), \int_R h(x-t)g(x)dx \right\rangle$$

$$\int_R \int_R f(x)h(t-x)dx\, g(t)dt = \int_R f(t) \int_R h(x-t)g(x)dx\, dt.$$

Now, since $f(t)$, $g(t)$, and $h(t)$ are Lebesgue integrable, we can interchange the order of integration to obtain

$$\int_R \int_R f(x)h(t-x)g(t)\,dx\,dt = \int_R \int_R f(t)h(x-t)g(x)\,dt\,dx.$$

Interchanging the dummy integration variables x and t on either side of the above equation shows that the definition of convolution as per equation (7.27) is consistent with that presented in Chapter 6 for absolutely integrable functions.

To examine the existence of the convolution of two distributions, we first consider the simple product of a distribution $f(t)$ and a function $h(t) \in D$. From Chapter 4 we know

$$\langle f(t)h(t), g(t) \rangle = \langle f(t), h(t)g(t) \rangle.$$

Now, since $h(t) \in D$ and $g(t) \in S$, we have $h(t)g(t) \in S$ (see Problem 4), and consequently the above product is valid and exists. In other words, the product of a distribution and a function will always exist if the function is from the space D.

Now that we have established existence conditions for the product of a distribution and a function, let us consider the following mapping:

$$\langle f(t)h(t), g(t) \rangle = \langle f(t), h(t)g(t) \rangle.$$

Fourier transforming both sides of the above equation yields

$$\langle \mathcal{F}[f(t)h(t)], G(-w) \rangle = \langle F(w), \mathcal{F}[h(t)g(t)] \rangle.$$

However, since both $h(t)$ and $g(t)$ are functions, we have

$$\mathcal{F}[h(t)g(t)] = H(w) * G(w),$$

or

$$\langle \mathcal{F}[f(t)h(t)], G(-w) \rangle = \langle F(w), H(-w) * G(-w) \rangle.$$

However, application of equation (7.27) on the right-hand side of the above equation gives

$$\langle \mathcal{F}[f(t)h(t)], G(-w) \rangle = \langle F(w) * H(w), G(-w) \rangle,$$

or

$$\mathcal{F}[f(t)h(t)] = F(w) * H(w). \tag{7.28}$$

In equation (7.28) both $F(w)$ and $H(w)$ are considered, in the broader sense, as distributions. This equation tells us when the convolution of two distributions exists. In other words, the convolution of $H(w)$ and $F(w)$ exists if the product function $f(t)h(t)$ exists. We know, however, by the way we set up this product that since $h(t) \in D$ it does indeed exist.

Because of the dual nature of the Fourier transform (Theorem 7.5) and the commutative property of the convolution product, we can generalize the above results to the time domain. That is to say, the convolution of two distributions $f(t)$ and $h(t)$ exists if either $\mathcal{F}[f(t)] = F(w) \in D$ or $\mathcal{F}[h(t)] = H(w) \in D$. We summarize the previous development with the following three theorems.

Theorem 7.12. Let the distributions $f(t)$ and $h(t)$ have Fourier transforms given as $F(w)$ and $H(w)$, respectively. Then the convolution product $f(t) * h(t)$ exists if either one or both of the transforms are functions in the space D.

We note here that Theorem 7.12 presents sufficient, but not necessary, conditions for existence.

Theorem 7.13 (Convolution). Let the distributions $f(t)$ and $h(t)$ possess Fourier transforms given as $F(w)$ and $H(w)$, respectively. If $F(w) \in D$ and/or $H(w) \in D$, then the Fourier transform of the convolution product $f(t) * h(t)$ exists and is given as the simple product of the transforms, that is,

$$\mathcal{F}[f(t) * h(t)] = F(w)H(w).$$

Theorem 7.14 (Product). Let the distributions $f(t)$ and $F(w)$ be Fourier transform pairs. Also, let $h(t) \in D$ have a Fourier transform given by $H(w)$. Then the Fourier transform of the simple product $f(t)h(t)$ exists and is given as the convolution product of the transforms, that is,

$$\mathcal{F}[f(t)h(t)] = F(w) * H(w).$$

Example 11. We now demonstrate that the delta distribution is the unit element under convolution. That is to say,

$$\delta(t) * f(t) = f(t).$$

This is easily demonstrated using Theorem 7.13, that is,

$$\mathcal{F}[\delta(t) * f(t)] = 1F(W) = F(w).$$

Thus, taking the inverse Fourier transform of both sides of the above equation, we find

$$\delta(t) * f(t) = f(t). \tag{7.29}$$

Example 12. We next consider convolution with the shifted delta distribution $\delta(t - a)$:

$$\mathcal{F}[\delta(t - a) * f(t)] = 1e^{-2\pi i w a}F(w) = F(w)e^{-2\pi i w a}.$$

Again, inverse Fourier transforming both sides of the above equation yields our desired results, that is,

$$\delta(t - a) * f(t) = f(t - a). \tag{7.30}$$

A graphical illustration of equation (7.30) is presented in Figure 7.5. In this figure we see that when we convolve a function (or distribution) $f(t)$ with a

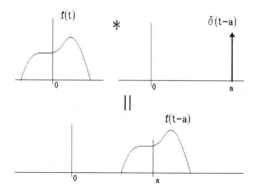

Figure 7.5 Graphical illustration of convolution of a shifted delta distribution and a function.

shifted delta distribution located at $t = a$, we displace or move the function to the location $t = a$.

Example 13. Let us now demonstrate that the unit constant function $u(t) = 1$ can be convolved with a pulse function. Inasmuch as $\mathcal{F}[p_a(t)] = 2a\operatorname{sinc}(2\pi wa) \in D$, this convolution exists. Furthermore,

$$\mathcal{F}[u(t) * p_a(t)] = 2a\delta(w)\operatorname{sinc}(2\pi wa) = 2a\operatorname{sinc}(0)\delta(w) = 2a\delta(w).$$

Inverse transforming both sides of the above equation yields

$$u(t) * p_a(t) = 2a.$$

Example 14. Let us now consider the convolution of the unit step function with itself, that is,

$$u(t) * u(t).$$

We first note that $\mathcal{F}[u(t)] = \delta(w)$, and thus we see that the Fourier transform of this function is not a function in the space D. Consequently, Theorem 7.12 cannot be used to guarantee the existence of the convolution product. However, since this theorem only provides sufficient conditions, we cannot use the theorem to guarantee that the convolution does not exist. However, we do note

$$u(t) * u(t) = \int_R u(t - x)u(x)\,dx = \int_R dx,$$

which does not exist.

PHYSICAL INTERPRETATION OF CONVOLUTION

In Chapter 6 when we discussed convolution of two functions we presented a graphical interpretation of this convolution product. Also in that chapter we pointed out how important convolution was to the physical application of Fourier analysis. Because of that importance we now present a brief physical, or systems, interpretation of convolution with the intent of shedding additional light on the subject. To accomplish this we must first define what we mean by a system.

In general terms, *a system is defined to be a mapping of a set of input functions to a set of output functions.* An equivalent, but less compact, description is: *A system is defined as a mathematical abstraction devised to serve as a model for physical phenomenon. It consists of an input function $f(t)$, an output function $y(t)$, and a cause–effect relationship between them.* The following notation is often used to denote this cause–effect relationship:

$$y(t) = S[f(t)]. \tag{7.31}$$

We now limit our attention to single-input–single-output systems that are linear and invariant. We say a system is *linear* if and only if for

$$S[f_1(t)] = y_1(t) \quad \text{and} \quad S[f_2(t)] = y_2(t)$$

we have

$$S[af_1(t) + bf_2(t)] = aS[f_1(t)] + bS[f_2(t)] = ay_1(t) + by_2(t).$$

where a and b are arbitrary constants. A system is said to be *invariant* (or *stationary*) if and only if for any value τ we have

$$S[f(t - \tau)] = y(t - \tau).$$

In words, a system is invariant if a displaced or shifted input function $f(t - \tau)$ yields the output function with the same displacement or shift $y(t - \tau)$.

We now demonstrate the fact that a linear invariant system can be completely characterized by a knowledge of how it responds when the input function is an impulse or delta "function." Let S be a system that maps a general input function $f(t)$ to the corresponding output function $y(t)$. The *impulse response* of such a system, denoted as $h(t)$, is defined as the output that results when the system is excited with a delta "function," that is,

$$h(t) = S[\delta(t)].$$

By assumption, S is both linear and invariant, and consequently an input of the form $a\delta(t) + b\delta(t - \tau)$ will result in the output function $ah(t) + bh(t - \tau)$. Shown in Figure 7.6 is an arbitrary function $f(\Delta T)$ whose domain has been divided into equal intervals of width ΔT. As indicated in this figure, at every location $k\Delta T$ a rectangle of height $f(k\Delta T)$ and width ΔT is constructed. Now, as input to our system S, let us consider a linear combination of shifted impulses $\delta(t - k\Delta T)$, each of which is weighted by the constant $f(k\Delta T)\Delta T$, that is,

$$g(t) = \sum_{k=-\infty}^{\infty} f(k\Delta T)\Delta T \delta(t - \Delta T), \tag{7.32}$$

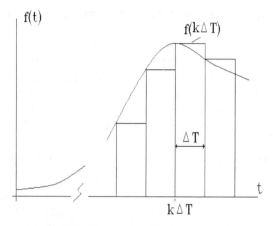

Figure 7.6 Arbitrary function for convolution.

Again, linearity and invariance of the system guarantee that the output must be the same linear combination of the shifted impulse response functions:

$$y(t) = \sum_{k=-\infty}^{\infty} f(k\Delta T)\Delta Th(t - \Delta T). \tag{7.33}$$

Essentially, what we have is an infinite train of weighted impulses exciting the system, with the resulting output being a linear combination of the responses to each of the impulses. If now in equations (7.32) and (7.33) we "shrink" the interval width ΔT to zero and use a somewhat heuristic calculus approach, we can reason that $k\Delta T$ approaches a continuous variable x, ΔT becomes dx, and the summation goes to an integral. Equation (7.32) then becomes

$$g(t) = \int_{-\infty}^{\infty} f(x)\delta(t - x)dx = f(t) * \delta(t) = f(t).$$

We see, therefore, that (in the limit) our input train of weighted impulses equals the general function $f(t)$. Using similar reasoning, we demonstrate that equation (7.33) becomes

$$y(t) = \int_{-\infty}^{\infty} f(x)h(t - x)dx = f(t) * h(t). \tag{7.34}$$

Equation (7.34) tells us that the output of a system $y(t)$, resulting from any general input function $f(t)$, is simply the convolution of that input function with the impulse response function of the system. Therefore, if we know how a system responds when excited by an impulse, then we can use equation (7.34) to calculate its response to any general input function.

FOURIER TRANSFORM OF A PERIODIC FUNCTION: THE FOURIER SERIES

In this section we demonstrate the fact that the Fourier transform of a periodic function is a sequence whose terms are the complex Fourier series coefficients of the function. Before we begin, however, we must first take a brief detour and discuss the convolution product and the simple product of a function with a comb function.[†] We begin with the simple product. By definition, we have

$$\langle f(t)\text{comb}_T(t),g(t)\rangle = \langle f(t)\sum_{k=-\infty}^{\infty}\delta(t-kT),g(t)\rangle$$

$$= \left\langle \sum_{k=-\infty}^{\infty}\delta(t-kT),f(t)g(t)\right\rangle$$

$$= \{f(kT)g(kT)\}.$$

$$\langle f(t)\text{comb}_T(t),g(t)\rangle = \left\langle \sum_{k=-\infty}^{\infty}f(kT)\delta(t-kT),g(t)\right\rangle.$$

Thus,

$$f(t)\text{comb}_T(t) = \sum_{k=-\infty}^{\infty}f(kT)\delta(t-kT) = \{f(kT)\}.$$

In other words, the product of a function $f(t)$ and a comb function of spacing T is a sequence whose terms are the values of the function at the discrete locations $t = kT$, $k = -\infty,\ldots,-1,0,1,\ldots,\infty$.

We next consider the convolution of a comb function and a function $f(t)$ which has bounded support. When we say a function has *bounded support* on the interval I we mean that the function is equal to zero outside that interval I. As we saw in a previous section, if we convolve a function $f(t)$ with a shifted delta function located at $t = a$, then we move or displace that function to the location $t = a$ (see Figure 7.5). Consequently, if we convolve a function $f(t)$ with a comb function of spacing T, then we essentially copy the function to locations $t = kT$, $k = -\infty,\ldots,-1,0,1,\ldots,\infty$. This is graphically illustrated in Figure 7.7. Obviously, if the function $f(t)$ has bounded support over the interval $T = [-T/2,T/2]$, then these "copied" functions will not overlap each other. One interpretation of this is: When we convolve a function of bounded support (on the interval T) with a comb function of width T, then we convert that function to a periodic function. Let us denote this periodic function as $g(t)$, that is,

$$g(t) = f(t)*\text{comb}_T(t). \tag{7.35}$$

[†]In this section we again relax our terminology and refer to both the delta and comb distributions as the delta and comb functions.

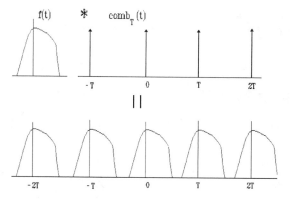

Figure 7.7 Convolution of function of bounded support with comb function to obtain a periodic function.

We next take the Fourier transform of both sides of the above equation and use the convolution theorem (Theorem 7.13) along with the fact that the transform of $\text{comb}_T(t)$ is $(1/T)\text{comb}_{1/T}(w)$ to obtain

$$G(w) = \frac{1}{T}F(w)\text{comb}_{1/T}(w)$$

or

$$G(w) = \frac{1}{T}\left\{F\left(\frac{j}{T}\right)\right\}. \qquad (7.36)$$

Thus we see that the Fourier transform of this function is a sequence whose values are obtained from the Fourier transform of the function $f(t)$ at discrete frequency locations $w = j/T$, $j = -\infty,\ldots,-1,0,1,\ldots,\infty$.

Using the definition of the Fourier transform (equation 6.1), we have

$$F(w) = \int_{-\infty}^{\infty} f(t)e^{-2\pi i w t}\,dt = \int_{-T/2}^{T/2} f(t)e^{-2\pi i w t}\,dt.$$

If we now let $w = j/T$ in the above equation and divide both sides by $1/T$, we obtain

$$\frac{F(j/T)}{T} = \frac{1}{T}\int_{-T/2}^{T/2} f(t)e^{-2\pi i j t/T}\,dt.$$

Referring back to Chapter 5 (equation (5.7)), we see that the integral on the right-hand side of the above equation is the expression for the complex Fourier series coefficients of the function $f(t)$. Thus,

$$C_j = \frac{F(j/T)}{T}. \qquad (7.37)$$

In other words, when we take the Fourier transform of a periodic function we obtain a sequence of terms that turn out to be the complex Fourier series coefficients of the periodic function.

Example 15. The boxcar function shown in Figure 1.3 can be described as the convolution of a pulse function of half-width $a = 1/2$ and a comb function with spacing $T = 2$, that is,

$$g(t) = p_{1/2}(t) * \text{comb}_2(t).$$

The Fourier transform of this pulse function is given as

$$F(w) = \text{sinc}(\pi w),$$

from which it follows that

$$C_j = \frac{1}{2}\text{sinc}\left(\frac{j\pi}{2}\right).$$

Therefore the complex Fourier series representation of this function is given as (see equation (5.6))

$$f(t) = \sum_{j=-\infty}^{\infty} C_j e^{\pi i j t}.$$

or

$$f(t) = \sum_{j=-\infty}^{\infty} \frac{1}{2}\text{sinc}\left(\frac{j\pi}{2}\right)[\cos(j\pi t) + i\sin(j\pi t)].$$

Now, splitting the summation from $k = -\infty$ to -1 and $k = +1$ to ∞, we obtain

$$f(t) = \frac{1}{2}\text{sinc}(0) + \sum_{j=-1}^{-\infty} \frac{1}{2}\text{sinc}\left(\frac{j\pi}{2}\right)[\cos(j\pi t) + i\sin(j\pi t)]$$

$$+ \sum_{j=1}^{\infty} \frac{1}{2}\text{sinc}\left(\frac{j\pi}{2}\right)[\cos(j\pi t) + i\sin(j\pi t)].$$

Using the fact that both the sinc and cosine are even functions and that the sine is an odd function, we substitute $-j$ for j in the first summation (and change the summation limits appropriately) to obtain

$$f(t) = \frac{1}{2} + \sum_{j=1}^{\infty} \text{sinc}\left(\frac{j\pi}{2}\right)\cos(j\pi t),$$

which agrees with the results obtained in Chapter 1.

SUMMARY

In this chapter we introduced the concept of the Fourier transform of a distribution. We showed that this transform possessed all of the same properties as the Fourier transform of an absolutely integrable function that was discussed in Chapter 6. As a matter of fact, distributions can be considered a larger class of functions that contain the absolutely integrable and Lebesgue integrable functions as a subset.

This distribution theory approach to Fourier analysis opened the door and allowed us to consider the Fourier transform of many more functions. These new functions included the well-behaved sine, cosine, and sinc functions as well as the "maverick" ones such as the delta and comb functions.

Convolution was again discussed in this chapter in light of distribution theory which allowed us to consider the convolution product of a much wider (and perhaps more useful) class of functions than those presented in Chapter 6. This approach also enabled us to present sufficient conditions for the existence of the convolution product of two functions.

As in the previous chapter, our theoretical results were punctuated with examples to illustrate their application and usefulness.

PROBLEMS

1 Use Theorems 6.9 and 6.12 to demonstrate the fact that if $g(t) \in S$, then $\mathcal{F}[g(t)] \in S$.

2 Show that the chain mapping of the distribution Fourier transform (see Figure 7.1) is both linear and continuous.

3 Prove parts (b), (c), and (d) of Theorem 7.11. [Use the proof of part (a) as a guide].

4 Use the definitions of the functions in the spaces D and S (see Chapter 4) to demonstrate that if $h(t) \in D$ and $g(t) \in S$, then the product function $h(t)f(t)$ is an element of the space S.

5 Determine the Fourier transform of the function that is the sum of two shifted pulse functions, that is, $g(t) = p_a(t - b) + p_a(t + b)$.

6 What is the Fourier transform of the derivative of the function described in problem 5?

7 Use equation (7.10), the derivative theorem (Theorem 7.8), and the fact that the derivative of $\sin(2\pi at)$ is $2\pi a \cos(2\pi at)$ to determine the Fourier transform of $\cos(2\pi at)$. Check your answer against equation (7.9).

8 Use the derivative theorem (Theorem 7.6) to determine the Fourier transform of $t \sin(2\pi at)$.

9 What is the Fourier transform of $[t \cos(t) - \sin(t)]/t^2$? [*Hint*: Consider the derivative of $\mathrm{sinc}(t)$].

10 Use the results of Theorem 7.12 to determine if we can guarantee the existence of the convolution of the following functions:

(a) $f(t) = 1$ and $g(t) = \exp[-t^2]$,

(b) $f(t) = p_a(t)$ and $g(t) = \exp[-t^2]$,

(c) $f(t) = \delta(t)$ and $g(t) = \delta(t - a)$,

(d) $f(t) = \text{comb}_T(t)$ and $g(t) = p_a(t) * p_a(t)$,

(e) $f(t) = \text{comb}_T(t)$ and $g(t) = \text{comb}_T(t)$.

11   Use equation (7.37) and the convolution of two pulse functions to determine the Fourier series representation of the triangle function given by equation (5.56) (see Figure 5.6a). Check your answer with that given in Chapter 5, Example 1.

12   Use the first shifting theorem (Theorem 7.8) and the results of problem 11 to determine the Fourier series representation of the "V" function given by equation (5.58) and shown in Figure 5.8a. Check your results with that given in Chapter 5, Example 3.

13   Consider the function $f(t) = t^2$ over the interval $[-T/2, T/2]$ and zero elsewhere. What is $\mathcal{F}[f(t)]$? Using this result, determine the Fourier series representation of the function shown in Figure 5.9a. Check your results with those given in Chapter 5, Example 4.

14   From equation (7.9) (and Figure 7.2a) we know that the Fourier transform of a pure cosine wave is given by two equal delta functions (of strength $\frac{1}{2}$) located at $w = +a$ and $w = -a$. In real life a pure cosine signal cannot be physically realized since it cannot be observed (or recorded) for all time between $-\infty$ and ∞. Instead, we have $g(t) = p_T(t)\cos(2\pi at)$, where T is the observation period. What does the Fourier transform of $g(t)$ look like?

15   Use the results of problem 14 and equation (7.11) to determine the Fourier series representation of the function shown in Figure 5.11a.

16   Suppose you are given the function

$$g(t) = f_1(t)\cos(2\pi w_0 t) + f_2(t)\cos(4\pi w_0 t).$$

How would you recover the functions $f_1(t)$ and $f_2(t)$ from $g(t)$? What restrictions must be placed on $f_1(t)$, $f_2(t)$, and w_0 for your solution to work?

BIBLIOGRAPHY

Arsac, J., *Fourier Transforms and the Theory of Distributions*, Prentice-Hall, Englewood Cliffs, N.J., 1966.

Weaver, H. J., *Applications of Discrete and Continuous Fourier Analysis*, John Wiley & Sons, New York, 1983.

8

THE DISCRETE FOURIER TRANSFORM

Thus far in this text, we have dealt with the Fourier series representation and Fourier transform of a function or distribution. Both the transform and series may be considered as a measure of the frequency content of the involved functions. Also, by their very definition, both require the evaluation of integral expressions. Therefore, in a strict sense, we were only able to determine the Fourier transform (or series) of functions which could be described by an analytical (or formula) expression; even then, these expressions had to be relatively uncomplicated in order to evaluate the associated integrals.

In a practical sense, we are more often than not dealing with functions, or signals, which are produced by some physical or experimental measurement. For example, we may be required to determine the Fourier transform of the signal produced by an accelerometer mounted on a structure. These signals will almost never have an analytical expression, and therefore they must be analyzed by digitizing them and performing the required mathematical manipulation on the resulting sequence using a digital computer.

In this chapter we discuss finite sequences and the *discrete Fourier transform*, which is an operation that maps one sequence to another.

NTH-ORDER SEQUENCES

A finite sequence of N terms, or *Nth-order sequence*, is defined as a function whose domain is the set of integers $\{0, 1, 2, ..., N-1\}$ and whose range is the set of terms $\{f(0), f(1), f(2), ..., f(N-1)\}$. Formally, an Nth-order sequence is the set of ordered pairs

$$\{[0, f(0)], [1, f(1)], [2, f(2)], ..., [N-1, f(N-1)]\}.$$

TABLE 8.1. $f(k) = 1/(k+1)$

k	$f(k)$
0	1.0000
1	0.5000
2	0.3333
3	0.2500
4	0.2000
5	0.1666
6	0.1429
7	0.1250

In this chapter we use a shorter notation and follow the common practice of denoting the sequence as $\{f(k)\}$ and the kth term of the sequence as $f(k)$. To clarify these remarks let us consider the sequence $\{f(k)\}$ whose terms are defined as

$$f(k) = 1/(k+1), \qquad k = 0,1,...,N-1.$$

Thus,

$$\{f(k)\} = \{1, 1/2, 1/3, ..., 1/N\}.$$

For convenience, in this chapter we often show the sequence in both table and graphical form. For example, the terms of this previous sequence (for $N = 8$) are given in Table 8.1 and are illustrated graphically in Figure 8.1.

In the previous example, the terms of the Nth-order sequence were indeed generated by a mathematical formula. However, as already noted, this will not be the case in general.

We now present some basic algebraic rules for sequences. The *sum of two Nth-order sequences* $\{f(k)\}$ and $\{g(k)\}$, denoted as $\{f(k)+g(k)\}$, is defined as the sequence whose terms are given by $f(k)+g(k)$ for $k \in [0, N-1]$. In other words, we add the individual terms, for example,

$$\{0,1,2\} + \{4,2,1\} = \{4,3,3\}.$$

Subtraction is the inverse of addition. That is, to subtract two sequences, we subtract the individual terms. Note that addition and subtraction are only defined for sequences that have the same order.

The *product of two sequences* $\{f(k)\}$ and $\{g(k)\}$, denoted as $\{f(k)g(k)\}$, is defined as the sequence whose terms are given by $f(k)g(k)$ for $k \in [0, N-1]$. In other words, we multiply the individual terms, for example,

$$\{0,1,2\}\{4,2,1\} = \{0,2,2\}.$$

Division is the inverse of multiplication; thus, to divide the sequence $\{f(k)\}$ by $\{g(k)\}$, we divide the individual terms:

$$\frac{f(k)}{g(k)}, \qquad g(k) \neq 0, \qquad k \in [0, N-1].$$

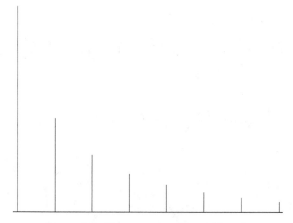

Figure 8.1 The sequence $f(k) = 1/(k+1)$.

Note. In general, the terms of the sequences may be complex, and therefore the algebraic rules for multiplication (and division) of complex numbers must be used.

To multiply a sequence by a constant (perhaps complex), we simply multiply each term of the sequence by the constant. For example,

$$3\{1,2,3\} = \{3,6,9\}.$$

Finally, we say a sequence is *bounded* if and only if all its terms are finite valued.

THE DISCRETE FOURIER TRANSFORM

Given any bounded Nth-order sequence $\{f(k)\}$, the *discrete Fourier transform pair* is defined as

$$F(j) = \frac{1}{N}\sum_{k=0}^{N-1} f(k)e^{-2\pi i k j/N}, \qquad j \in [0, N-1], \tag{8.1}$$

$$f(k) = \sum_{j=0}^{N-1} F(j)e^{2\pi i k j/N}, \qquad k \in [0, N-1]. \tag{8.2}$$

Equation (8.1) is called the *(direct) discrete Fourier transform*, and equation (8.2) is known as the *inverse discrete Fourier transform*.

In the most general situation, both sequences $\{f(k)\}$ and $\{F(j)\}$ will be complex, that is,

$$f(k) = f_R(k) + if_I(k)$$

and

$$F(j) = F_R(j) + iF_I(j).$$

In this situation equation (8.1) becomes

$$F(j) = \frac{1}{N} \sum_{k=0}^{N-1} [f_R(k) + if_I(k)] \left[\cos\left(\frac{2\pi k j}{N}\right) - i \sin\left(\frac{2\pi k j}{N}\right) \right], \quad j \in [0, N-1]$$

or

$$F_R(j) = \frac{1}{N} \sum_{k=0}^{N-1} \left[f_R(k) \cos\left(\frac{2\pi k j}{N}\right) + f_I(k) \sin\left(\frac{2\pi k j}{N}\right) \right], \quad j \in [0, N-1],$$

$$(8.3)$$

$$F_I(j) = \frac{1}{N} \sum_{k=0}^{N-1} \left[f_I(k) \cos\left(\frac{2\pi k j}{N}\right) - f_R(k) \sin\left(\frac{2\pi k j}{N}\right) \right], \quad j \in [0, N-1].$$

Similarly, we are able to rewrite equation (8.2) in component form as

$$f_R(k) = \sum_{j=0}^{N-1} \left[F_R(j) \cos\left(\frac{2\pi k j}{N}\right) - F_I(j) \sin\left(\frac{2\pi k j}{N}\right) \right], \quad k \in [0, N-1],$$

$$(8.4)$$

$$f_I(k) = \sum_{j=0}^{N-1} \left[F_I(j) \cos\left(\frac{2\pi k j}{N}\right) + F_R(j) \sin\left(\frac{2\pi k j}{N}\right) \right], \quad k \in [0, N-1].$$

When actually performing the discrete Fourier transform or the inverse discrete Fourier tranform of a sequence, equations (8.3) and (8.4) are used. However, for the sake of notational simplicity and mathematical manipulational purposes, we often use a simplified form. If we define the *weighting kernel* W_N as

$$W_N = e^{2\pi i/N},$$

then equations (8.1) and (8.2) become

$$F(j) = \frac{1}{N} \sum_{k=0}^{N-1} f(k) W_N^{-kj}, \quad j \in [0, N-1], \qquad (8.5)$$

$$f(k) = \sum_{j=0}^{N-1} F(j) W_N^{kj}, \quad k \in [0, N-1]. \qquad (8.6)$$

As we can appreciate from the previous equations, the discrete Fourier transform is an operation that maps an Nth-order sequence $\{f(k)\}$ to another Nth-order sequence $\{F(j)\}$. For example, the second-order sequence $\{f(k)\} = \{1, 2\}$ has a discrete Fourier transform given by

$$F(0) = (1/2)(f(0)[\cos(0) - i\sin(0)] + f(1)[\cos(0) - i\sin(0)]),$$

$$F(1) = (1/2)(f(0)[\cos(0) - i\sin(0)] + f(1)[\cos(\pi) - i\sin(\pi)]),$$

or

$$F(0) = (1/2)[1 + 2] = 3/2,$$
$$F(1) = (1/2)[1 - 2] = -1/2.$$

Thus,

$$\{F(j)\} = (3/2, -1/2)\}.$$

Using the inverse transform operation on this sequence, we obtain

$$f(0) = F(0)[\cos(0) + i\sin(0)] + F(1)[\cos(0) + i\sin(0)],$$
$$f(1) = F(0)[\cos(0) + i\sin(0)] + F(1)[\cos(\pi) + i\sin(\pi)],$$

or

$$f(0) = 3/2 + (-1/2) = 1,$$
$$f(1) = 3/2 + (-1)(-1/2) = 2.$$

Thus,

$$\{f(k)\} = \{1, 2\}.$$

Using a digital computer we are able to perform the following.

Example 1. Consider the 16th-order sequence given in Table 8.2. Application of equation (8.3) (on a digital computer) yields the 16th-order sequence given in Table 8.3. Similarly, using equation (8.4) on the data in Table 8.3 will return the original sequence presented in Table 8.2. The sequence presented in Table 8.2 is illustrated graphically in Figure 8.2, and the magnitude of the complex data presented in Table 8.3 is illustrated in Figure 8.3. The reader is urged to verify these calculations using a digital computer and equations (8.3) and (8.4) (see problem 1).

In the previous illustrations we have demonstrated (for two particular sequences) that the discrete Fourier transform possesses complete reciprocity. As it turns out, this result is true in general. That is to say, if we obtain the sequence $\{F(j)\}$ from the sequence $\{f(k)\}$ using equation (8.5), then equation (8.6) will uniquely return $\{f(k)\}$ from $\{F(j)\}$. We soon demonstrate this result, but first we require the results of the following.

Lemma 8.1. If m is any integer not equal to 0 or N, then

$$\sum_{j=0}^{N-1} W_N^{mj} = 0.$$

Proof. We begin by constructing the sequence $\{V(j)\}$ whose terms are given by

$$V(j) = \frac{W_N^{mj}}{W_N^m - 1}.$$

TABLE 8.2 Example 1: $\{f(k)\}$

k	$f(k)$
0	1.00000
1	.39063
2	.20661
3	.12755
4	.08651
5	.06250
6	.00726
7	.03698
8	.02973
9	.02441
10	.02041
11	.01731
12	.01487
13	.01291
14	.01132
15	.01000

TABLE 8.3 Example 1: $\{F(j)\}$

j	Real	Imaginary
0	.13119	.00000
1	.08832	−.03293
2	.06871	−.02986
3	.05861	−.02422
4	.05284	−.01866
5	.04939	−.01354
6	.04734	−.00881
7	.04624	−.00434
8	.04590	.00000
9	.04624	.00434
10	.04734	.00881
11	.04939	.01354
12	.05284	.01866
13	.05861	.02422
14	.06871	.02986
15	.08832	.03293

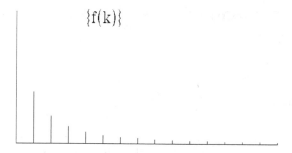

Figure 8.2 Example 1: "Time domain" sequence {f(k)}.

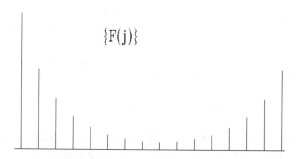

Figure 8.3 Example 1: "Frequency domain" sequence {|F(j)|}.

From the above equation it follows that

$$V(j+1) - V(j) = \frac{W_N^{m(j+1)} - W_N^{mj}}{W_N^m - 1},$$

$$V(j+1) - V(j) = \frac{W_N^{mj}(W_N^m - 1)}{W_N^m - 1} = W_N^{mj}.$$

Thus,

$$\sum_{j=0}^{N-1} W_N^{mj} = \sum_{j=0}^{N-1} V(j+1) - V(j) = V(N) - V(0),$$

$$\sum_{j=0}^{N-1} W_N^{mj} = \frac{W_N^{mN} - W_N^{m0}}{W_N^m - 1} = \frac{e^{2\pi i m} - e^0}{W_N^m - 1}$$

$$\sum_{j=0}^{N-1} W_N^{mj} = \frac{1 - 1}{W_N^m - 1} = 0. \qquad \text{Q.E.D.}$$

We are now in a position to prove the following.

Theorem 8.1 (Reciprocity). The discrete Fourier transform possesses complete reciprocity; that is to say, it is unique.

Proof. We begin by writing

$$h(k) = \sum_{j=0}^{N-1} F(j) W_N^{kj}, \qquad k \in [0, N-1],$$

$$h(k) = \sum_{j=0}^{N-1} \left[\frac{1}{N} \sum_{l=0}^{N-1} f(l) W_N^{-lj} \right] W_N^{kj}, \qquad k \in [0, N-1],$$

$$h(k) = \frac{1}{N} \sum_{l=0}^{N-1} f(l) \left[\sum_{j=0}^{N-1} W_N^{(k-l)j} \right], \qquad k \in [0, N-1].$$

However, from Lemma 8.1 we know that in the above equation the summation in brackets is zero unless $l = k$, in which case it is equal to N. Thus,

$$h(k) = \frac{1}{N} f(k) N = f(k). \qquad \text{Q.E.D.}$$

The next two theorems deal with the concept of extending the domain of definition of a sequence and its discrete Fourier transform.

Theorem 8.2. The discrete Fourier transform and the discrete inverse Fourier transform are both periodic with periodicity N; that is, (a) $F(j + N) = F(j)$ and (b) $f(k + N) = f(k)$.

Proof.

(a) Using equation (8.1), we have

$$F(j + N) = \frac{1}{N} \sum_{k=0}^{N-1} f(k) e^{-2\pi i k(j+N)/N} = \frac{1}{N} \sum_{k=0}^{N-1} f(k) e^{-2\pi i kj/N} e^{-2\pi i k},$$

$$F(j + N) = \frac{1}{N} \sum_{k=0}^{N-1} f(k) e^{-2\pi i kj/N} = F(j).$$

(b) Using equation (8.2), we have

$$f(k + N) = \sum_{j=0}^{N-1} F(j) e^{2\pi i (k+N)j/N} = \sum_{j=0}^{N-1} F(j) e^{2\pi i kj/N} e^{2\pi i j},$$

$$f(k + N) = \sum_{j=0}^{N-1} F(j) e^{2\pi i kj/N} = f(k). \qquad \text{Q.E.D.}$$

Theorem 8.3. If the sequences $\{f(k)\}$ and $\{F(j)\}$ are discrete Fourier transform pairs, then:

(a) $F(-j) = F(N-j)$ and
(b) $f(-k) = f(N-k)$.

Proof.

(a) Using equation (8.1), we are able to write

$$F(N-j) = \frac{1}{N}\sum_{k=0}^{N-1} f(k)e^{-2\pi i k(N-j)/N} = \frac{1}{N}\sum_{k=0}^{N-1} f(k)e^{2\pi i kj/N}e^{-2\pi ik},$$

$$F(N-j) = \frac{1}{N}\sum_{k=0}^{N-1} f(k)e^{-2\pi i k(-j)/N} = F(-j).$$

(b) The proof of part (b) is similar and, consequently, is left as an exercise for the reader.
Q.E.D.

It is important to note that the discrete Fourier transform is defined over the domain $j \in [0, N-1]$ using values of $f(k)$ over the domain $k \in [0, N-1]$. However, the previous two theorems tell us that if we use equation (8.1) to calculate $F(j)$ for values of j outside the interval $[0, N-1]$, then we will still obtain a value *as if* the sequence were periodic. These remarks are illustrated graphically in Figure 8.4.

The previous theorems also allow us to interpret periodicity of the sequence $\{f(k)\}$. If we transform $\{f(k)\}$ to obtain $\{F(j)\}$ and then use equation (8.2) to recover $\{f(k)\}$ back from its transform $\{F(j)\}$, we can extend the sequence outside its domain. This is illustrated graphically in Figure 8.5.

It is extremely important to note that the previous remarks *in no way imply that the sequence* $\{f(k)\}$ *must be periodic* in order to calculate its discrete Fourier transform. As a matter of fact, $\{f(k)\}$ is an Nth-order sequence only defined over the finite interval $[0, N-1]$. However, if for some reason we have occasion to extend $\{f(k)\}$ outside this interval, we will use the previous theorems to do so.

PROPERTIES OF THE DISCRETE FOURIER TRANSFORM

Given any bounded sequence $\{f(k)\}$ we can always calculate its discrete Fourier transform using equation (8.1). This calculation requires N^2 complex mathematical multiplications and additions; for large values of N, this can become unwieldy. In this section we discuss several properties of the discrete Fourier transform that can often be used to simplify its calculation. In this section we should note a striking parallel between these properties and those

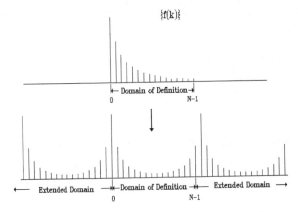

Figure 8.4   Periodicity of the discrete Fourier transform.

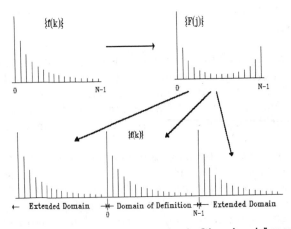

Figure 8.5   Concept of extended domain of a "time domain" sequence.

for the Fourier series and Fourier transform of a function (or distribution). We begin with the property of linearity.

Theorem 8.4 (Linearity). If $\{f(k)\}$ and $\{F(j)\}$ are discrete Fourier transform pairs as well as $\{g(k)\}$ and $\{G(j)\}$, then the discrete Fourier transform of the sequence $\{h(k)\} = a\{f(k)\} + b\{g(k)\}$ is given as the same linear combination of the respective transforms, that is,

$$\{H(j)\} = a\{F(j)\} + b\{G(j)\},$$

where $\{H(j)\}$ is the discrete Fourier transform of $\{h(k)\}$.

Proof. By definition, we have

$$H(j) = \frac{1}{N} \sum_{k=0}^{N-1} h(k) W_N^{-kj}, \qquad j \in [0, N-1],$$

$$H(j) = \frac{1}{N} \sum_{k=0}^{N-1} (af(k) + bg(k)) W_N^{-kj}, \qquad j \in [0, N-1],$$

$$H(j) = \frac{a}{N} \sum_{k=0}^{N-1} f(k) W_N^{-kj} + \frac{b}{N} \sum_{k=0}^{N-1} g(k) W_N^{-kj}, \qquad j \in [0, N-1],$$

$$H(j) = aF(j) + bG(j), \qquad j \in [0, N-1]. \qquad\qquad \text{Q.E.D.}$$

We now consider the shifting theorems.

Theorem 8.5 (First Shifting Theorem). If the discrete Fourier transform of the Nth-order sequence $\{f(k)\}$ is $\{F(j)\}$, then the discrete Fourier transform of the shifted sequence $\{f(k - n)\}$, $n \in [0, N-1]$, is given by $\{F(j)\} W_N^{-jn}\}$.

Proof. We begin with equation (8.6) and substitute $k - n$ for k to obtain

$$f(k - n) = \sum_{j=0}^{N-1} F(j) W_N^{(k-n)j} = \sum_{j=0}^{N-1} (F(j) W_N^{-jn}) W_N^{-kj}.$$

Thus, we see*

$$\{f(k - n)\} = \mathcal{F}^{-1}[\{F(j) W_N^{-jn}\}].$$

Now, taking the discrete Fourier transform of both sides of the above equation, we obtain our desired results:

$$\mathcal{F}[\{f(k - n)\}] = \mathcal{F}\mathcal{F}^{-1}[\{F(j) W_N^{-jn}\}] = \{F(j) W_N^{-jn}\}. \qquad \text{Q.E.D.}$$

Example 2. Consider the 16th-order sequence given in Table 8.4 and displayed graphically in Figure 8.6a. The discrete Fourier transform of this sequence is given in Table 8.5 and displayed graphically in Figure 8.6b. Let us now form a new sequence $\{g(k)\}$ by shifting this sequence to the right by 2. In other words, $\{g(k)\} = \{f(k - 2)\}$, which is given in Table 8.6 and shown graphically in Figure 8.7. A little explanation is in order at this point: When we shift each term to the right by any amount n, we naturally run into problems at the ends of the sequence interval. However, Theorems 8.2 and 8.3 indicate that if we need to extend the domain of the sequence $\{f(k)\}$, then it may be considered periodic. Thus, the terms that "fall off" on the right reappear on the left. Using the first shifting theorem (Theorem 8.5), we obtain the transform of

---

*Note: We use the same notation as in Chapter 6, that is, $F(j) = \mathcal{F}[\{f(k)\}]$ and $f(k) = \mathcal{F}^{-1}[\{F(j)\}]$ denote that $\{f(k)\}$ and $\{F(j)\}$ are discrete Fourier transform pairs.

TABLE 8.4   Example 2: $\{f(k)\}$

k	$f(k)$
0	1.00000
1	1.00000
2	1.00000
3	.00000
4	.00000
5	.00000
6	.00000
7	.00000
8	.00000
9	.00000
10	.00000
11	.00000
12	.00000
13	.00000
14	1.00000
15	1.00000

TABLE 8.5   Example 2: $\{F(j)\}$

j	Real	Imaginary
0	.31250	.00000
1	.26637	.00000
2	.15089	.00000
3	.02195	.00000
4	−.06250	.00000
5	−.07372	.00000
6	−.02589	.00000
7	.03540	.00000
8	.06250	.00000
9	.03540	.00000
10	−.02589	.00000
11	−.07372	.00000
12	−.06250	.00000
13	.02195	.00000
14	.15089	.00000
15	.26637	.00000

$\{g(k)\}$ from $\{f(k)\}$ with only 16 additional multiplications and additions (as opposed to 64 if we were to calculate it from scratch using equation (8.1)), that is,

$$G(j) = F(j)\left[\cos\left(\frac{\pi j}{4}\right) - i\sin\left(\frac{\pi j}{4}\right)\right].$$

TABLE 8.6   Example 2: $\{g(k)\}$

k	$g(k)$
0	1.00000
1	1.00000
2	1.00000
3	1.00000
4	1.00000
5	.00000
6	.00000
7	.00000
8	.00000
9	.00000
10	.00000
11	.00000
12	.00000
13	.00000
14	.00000
15	.00000

TABLE 8.7   Example 2: $\{G(j)\}$

j	Real	Imaginary
0	.31250	.00000
1	.18835	−.18835
2	.00000	−.15089
3	−.01552	−.01552
4	.06250	.00000
5	.05213	−.05213
6	.00000	−.02589
7	.02503	.02503
8	.06250	.00000
9	.02503	−.02503
10	.00000	.02589
11	.05213	.05213
12	.06250	.00000
13	−.01552	.01552
14	.00000	.15089
15	.18835	.18835

We now consider the following.

Theorem 8.6 (Second Shifting Theorem). If the discrete Fourier transform of the Nth-order sequence $\{f(k)\}$ is $\{F(j)\}$, then the discrete Fourier transform of the sequence $\{f(k)W_N^{nk}\}$ is given by $\{F(j-n)\}$, $n \in [0, N-1]$.

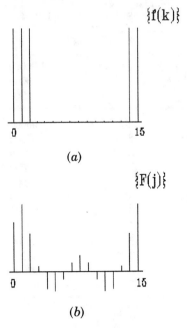

Figure 8.6   Example 2: "Pulse sequence" and its transform.

Proof. We begin with equation (8.5) and make the substitution $j - n$ for j, that is,

$$F(j - n) = \frac{1}{N} \sum_{k=0}^{N-1} f(k) W_N^{-k(j-n)} = \frac{1}{N} \sum_{k=0}^{N-1} [f(k) W_N^{nk}] W_N^{-kj}, \qquad j \in [0, N-1].$$

Comparing the above expression to equation (8.5), we see that $\{F(j - n)\}$ is indeed the discrete Fourier transform of $\{f(k) W_N^{nk}\}$.   Q.E.D.

Note the dual nature of the discrete Fourier transform illustrated by the previous two Theorems. Theorem 8.5 tells us that a "phase" shift in the k (or time) domain gives rise to a sinusoidal-type multiplication in the j (or frequency) domain. Theorem 8.6 tells us that a sinusoidal-type multiplication in the k domain gives rise to a "phase" shift in the j domain.

As a direct consequence of Theorem 8.6, we have the following theorem.

Theorem 8.7. If the sequences $\{f(k)\}$ and $\{F(j)\}$ are discrete Fourier transform pairs, then

(a) $\mathcal{F}[\{f(k)\cos(2\pi k n/N)\}] = (\frac{1}{2})[\{F(j + n) + F(j - n)\}],$
(b) $\mathcal{F}[\{f(k)\sin(2\pi k n/N)\}] = (i/2)[\{F(j + n) - F(j - n)\}],$
                for $n \in [0, N-1].$

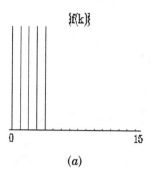

$$\{f(k)\}$$

0   15

(a)

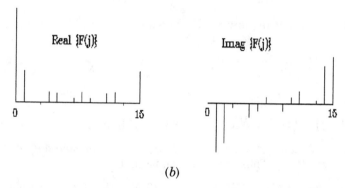

Real $\{F(j)\}$ Imag $\{F(j)\}$

0   15   0   15

(b)

Figure 8.7 Example 2: Shifted "pulse sequence" and its transform.

Proof. As a direct result of the previous theorem, we have

$$\mathcal{F}\left[\left\{f(k)\cos\left(\frac{2\pi k n}{N}\right) + if(k)\sin\left(\frac{2\pi k n}{N}\right)\right\}\right] = \{F(j-n)\},$$

$$\mathcal{F}\left[\left\{f(k)\cos\left(\frac{2\pi k n}{N}\right) - if(k)\sin\left(\frac{2\pi k n}{N}\right)\right\}\right] = \{F(j+n)\}.$$

Addition of these two equations yields the results of part (a), whereas subtraction yields part (b).   Q.E.D.

We now present a theorem concerning the discrete Fourier transform of the discrete Fourier transform of a sequence.

Theorem 8.8 (Transform of a Transform). If the Nth-order sequence $\{F(j)\}$ is the discrete Fourier transform of the sequence $\{f(k)\}$, then

$$\mathcal{F}[\mathcal{F}[\{f(k)\}]] = \frac{1}{N}\{f(-k)\} = \frac{1}{N}\{f(N-k)\}.$$

Proof. By assumption, we have

$$\mathcal{F}[\{f(k)\}] = \{F(j)\}.$$

Thus,

$$\mathcal{F}[\mathcal{F}[\{f(k)\}]] = \mathcal{F}[\{F(j)\}] = \frac{1}{N}\sum_{j=0}^{N-1}F(j)W_N^{-kj},$$

$$\mathcal{F}[\mathcal{F}[\{f(k)\}]] = \frac{1}{N}\sum_{j=0}^{N-1}F(j)W_N^{(-k)j} = \frac{1}{N}\{f(-k)\} = \frac{1}{N}\{f(N-k)\}.$$
$$\text{Q.E.D.}$$

Example 3. We define a delta sequence of order N as the Nth-order sequence

$$\{\delta(k)\} = \{1,0,...,0\}, \qquad k \in [0, N-1]. \tag{8.7}$$

Let us now consider the discrete Fourier transform of this sequence, that is,

$$\mathcal{F}[\{\delta(k)\}] = \frac{1}{N}\sum_{k=0}^{N-1}\delta(k)W_N^{-kj} = \frac{1}{N}\sum_{k=0}^{0}W_N^{-0j} = \frac{1}{N}, \qquad j \in [0, N-1].$$

Thus,

$$\mathcal{F}[\{\delta(k)\}] = \{1/N, 1/N,...,1/N\} = (1/N)\{1,1,...,1\}, \qquad j \in [0, N-1].$$

We call the Nth-order sequence $\{1,1,...,1\}$ the *unit constant sequence* and denote it as $\{U(k)\}$. Thus, the above equation becomes

$$\mathcal{F}[\{\delta(k)\}] = (1/N)\{U(j)\}, \qquad j \in [0, N-1], \quad k \in [0, N-1]. \tag{8.8}$$

Example 4. We now use Theorem 8.8 to determine the discrete Fourier transform of the unit constant sequence. From the previous example (equation 8.8) we know

$$(1/N)\{U(k)\} = \mathcal{F}[\{\delta(k)\}].$$

Now, taking the discrete Fourier transform of both sides of the above equation, we obtain

$$(1/N)\mathcal{F}[\{U(k)\}] = \mathcal{F}[\mathcal{F}[\{\delta(k)\}]] = (1/N)\{\delta(-k)\} = (1/N)\{\delta(k)\}.$$

Thus,

$$\mathcal{F}[\{U(k)\}] = \{\delta(j)\}. \tag{8.9}$$

Example 5. We now consider the discrete Fourier transform of the Nth-order sequence whose terms are given as

$$\cos\left(\frac{2\pi k n}{N}\right), \qquad k \in [0, N-1], \quad j \in [0, N-1].$$

We first consider the product of the unit constant sequence $\{U(k)\}$ and $\{\cos(2\pi k n/N)\}$, that is,

$$\{U(k)\}\left\{\cos\left(\frac{2\pi k n}{N}\right)\right\} = \left\{\cos\left(\frac{2\pi k n}{N}\right)\right\}.$$

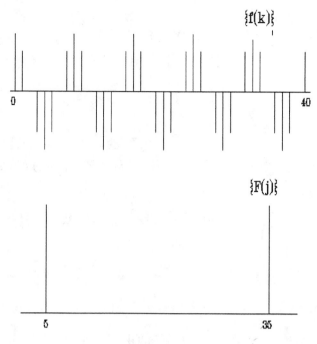

Figure 8.8   The sequence $\{f_k\} = \{\cos(2\pi kn/N)\}$ and its transform.

Taking the discrete Fourier transform of both sides of the above equation and using Theorem 8.7(a), we find

$$\mathcal{F}[\{\cos(2\pi k n/N)\}] = \mathcal{F}\left[\{U(k)\}\left\{\cos\left(\frac{2\pi k n}{N}\right)\right\}\right]$$

$$= \frac{1}{2}\{[\delta(j+n) + \delta(j-n)]\}.$$

Thus,

$$\mathcal{F}\left[\left\{\cos\left(\frac{2\pi k n}{N}\right)\right\}\right] = \frac{1}{2}\{[\delta(j+n) + \delta(j-n)]\}. \qquad (8.10)$$

This is illustrated graphically in Figure 8.8 for the 40th-order sequence $\{\cos(2\pi k n/N)\}$ with $n = 5$.

The *first forward difference* of a sequence is defined as

$$\{\Delta f(k)\} = \{f(k+1) - f(k)\}.$$

This forward difference of a sequence is analogous to the first derivative of a function. In Chapters 6 and 7 we presented theorems dealing with the derivative of the Fourier transform and the Fourier transform of the derivative of a function and a distribution. We now consider analogous results for the first difference of a sequence.

Theorem 8.9 (Difference of the Transform). If $\{F(j)\}$ is the discrete Fourier transform of the Nth-order sequence $\{f(k)\}$, then

$$\{\Delta F(j)\} = \mathcal{F}[\{f(k)(W_N^{-k} - 1)\}].$$

Proof. By definition, we have

$$\Delta F(j) = F(j+1) - F(j), \qquad j \in [0, N-1],$$

$$\Delta F(j) = \frac{1}{N} \sum_{k=0}^{N-1} f(k) W_N^{-k(j+1)} - \frac{1}{N} \sum_{k=0}^{N-1} f(k) W_N^{-kj}, \qquad j \in [0, N-1],$$

$$\Delta F(j) = \frac{1}{N} \sum_{k=0}^{N-1} [f(k) W_N^{-k} - f(k)] W_N^{-kj}, \qquad j \in [0, N-1],$$

$$\Delta F(j) = \frac{1}{N} \sum_{k=0}^{N-1} f(k)(W_N^{-k} - 1) W_N^{-kj}, \qquad j \in [0, N-1].$$

Comparing the above expression to equation (8.5), we obtain our desired results, that is,

$$\Delta F(j) = \mathcal{F}[\{f(k)(W_N^{-k} - 1)\}]. \qquad \text{Q.E.D.}$$

The following theorem is the analog of Theorem 8.9.

Theorem 8.10 (Transform of the Difference). If $\{F(j)\}$ is the discrete Fourier transform of the Nth-order sequence $\{f(k)\}$, then

$$\mathcal{F}[\{\Delta f(k)\}] = \{F(j)(W_N^j - 1)\} = \{F(j)\}\{W_N^j - 1\}.$$

Proof. Using equation (8.6), we have

$$\Delta f(k) = f(k+1) - f(k) = \sum_{j=0}^{N-1} F(j) W_N^{(k+1)j} - \sum_{j=0}^{N-1} F(j) W_N^{kj}, \quad k \in [0, N-1],$$

$$\Delta f(k) = \sum_{j=0}^{N-1} F(j)(W_N^j - 1) W_N^{kj}, \qquad k \in [0, N-1].$$

Thus, we see

$$\{\Delta f(k)\} = \mathcal{F}^{-1}[\{F(j)(W_N^j - 1)\}].$$

Now, taking the discrete Fourier transform of both sides of the above equation, we obtain our desired results, that is,

$$\mathcal{F}[\{\Delta f(k)\}] = \{F(j)(W_N^j - 1)\}. \qquad \text{Q.E.D.}$$

Example 6. Let us consider the Nth-order delta sequence

$$\{\delta(k)\} = \{1, 0, \ldots, 0, 0\}$$

and define a new sequence $\{f(k)\}$ as

$$\{f(k)\} = \{\Delta\{\delta(k)\} = \{-1, 0, \ldots, 0, 1\}.$$

However, $f(k)$ may be rewritten as the sum of two delta sequences, that is,

$$\{f(k)\} = -\{\delta(k)\} + \{\delta(k - n)\},$$

where $n = N - 1$.

We now determine the discrete Fourier transform of $\{f(k)\}$ using two different approaches. First we use the linearity theorem (Theorem 8.1) and the first shifting theorem (Theorem 8.5), that is,

$$\{F(j)\} = \mathcal{F}[-\{\delta(k)\}] + \mathcal{F}[\{\delta(k - n)\}],$$

$$\{F(j)\} = -1 + 1W_N^{-nj} = W_N^{-(N-1)j} - 1,$$

$$\{F(j)\} = W_N^{-N} W_N^{j} - 1 = W_N^{j} - 1.$$

Next we use Theorem 8.10 directly, that is,

$$\mathcal{F}[\{\Delta\delta(k)\}] = 1(W_N^{j} - 1) = W_N^{j} - 1.$$

SYMMETRY RELATIONS

In this section we discuss various forms of symmetry for the discrete Fourier transform. These results can be particularly useful for calculational purposes. We begin with the transform of the sequence $\{f(-k)\}$.

Theorem 8.11. If $\{F(j)\}$ is the discrete Fourier transform of the Nth-order sequence $\{f(k)\}$, then the discrete Fourier transform of $\{f(-k)\}$ is given as $\{F(-j)\}$.

Proof. Using Theorem 8.8, we know

$$\mathcal{F}[\mathcal{F}[\{f(k)\}]] = \mathcal{F}[\{F(j)\}] = (1/N)\{f(-k)\},$$

or

$$\mathcal{F}[\{F(j)\}] = (1/N)\{f(-k)\}.$$

Now, taking the transform of both sides of the above equation, we obtain

$$\mathcal{F}[\mathcal{F}[\{F(j)\}]] = (1/N)\{F(-j)\} = (1/N)\mathcal{F}[\{f(-k)\}].$$

Thus, canceling the $1/N$ terms in the above equation, we obtain our desired results, that is,

$$\{F(-j)\} = \mathcal{F}[\{f(-k)\}]. \qquad \text{Q.E.D.}$$

We now consider odd and even sequences. We say an Nth-order sequence $\{f(k)\}$ is *even* if and only if

$$f(N-k) = f(-k) = f(k), \qquad k \in [0, N-1].$$

Similarly, an Nth-order sequence $\{f(k)\}$ is said to be *odd* if and only if

$$f(N-k) = f(-k) = -f(k), \qquad k \in [0, N-1].$$

For example, the sequence in Table 8.4 is even because

$$f(1) = 1 = f(15),$$
$$f(2) = 1 = f(14),$$
$$f(3) = 0 = f(13),$$
$$f(4) = 0 = f(12),$$

etc.

The following theorem deals with the discrete Fourier transform of odd and even sequences.

Theorem 8.12. Assume that $\{F(j)\}$ is the discrete Fourier transform of the Nth-order sequence $\{f(k)\}$; then:

(a) If $\{f(k)\}$ is even, then $\{F(j)\}$ is even.
(b) If $\{f(k)\}$ is odd, then $\{F(j)\}$ is odd.

Proof.
(a) By assumption, $\{f(k)\}$ is even, which implies that $\{f(k)\} = \{f(-k)\}$. Thus,

$$\{F(j)\} = \mathcal{F}[\{f(k)\}] = \mathcal{F}[\{f(-k)\}].$$

However, Theorem 8.11 guarantees that $\mathcal{F}\{f(-k)\}] = \{F(-j)\}$; therefore,

$$\{F(j)\} = \{F(-j)\},$$

which implies that $\{F(j)\}$ is an even sequence.
(b) The proof of part (b) is analogous to that of part (a) and, consequently, is left as an exercise for the reader.   Q.E.D.

We note that the discrete Fourier transform of the even sequence presented in Table 8.4 is given in Table 8.5. Examination of this transform reveals that it is indeed even.

We say a sequence is *real* if and only if all of its terms are real. A sequence is said to be *imaginary* (or pure imaginary) if and only if all of its terms are imaginary. Given a sequence $\{f(k)\}$, its *complex conjugate sequence* [denoted as $\{f^*(k)\}]$ is defined as the sequence whose terms are the complex conjugates of those of $\{f(k)\}$. The next theorem deals with symmetry of complex sequences and their transforms.

Theorem 8.13. Assume that $\{F(j)\}$ is the discrete Fourier transform of the Nth-order sequence $\{f(k)\}$; then:

(a) If $\{f(k)\}$ is real, then $\{F(j)\} = \{F^*(-j)\} = \{F^*(N-j)\}$.
(b) If $\{F(j)\}$ is real, then $\{f(k)\} = \{f^*(-k)\} = \{f^*(N-k)\}$.

Proof.
   (a) By assumption, $\{f(k)\}$ is real, which implies $\{f^*(k)\} = \{f(k)\}$. Now, using equation (8.5), we write

$$F(j) = \frac{1}{N}\sum_{k=0}^{N-1} f(k)W_N^{-kj}, \qquad j \in [0, N-1].$$

Taking the complex conjugate of both sides of the above equation, we obtain

$$F^*(j) = \frac{1}{N}\sum_{k=0}^{N-1} f^*(k)W_N^{kj} = \frac{1}{N}\sum_{k=0}^{N-1} f(k)W_N^{kj}, \qquad j \in [0, N-1].$$

Finally, substitution of $-j$ for j in the above equation yields our desired results, that is,

$$F^*(-j) = \frac{1}{N}\sum_{k=0}^{N-1} f(k)W_N^{-kj} = F(j), \qquad j \in [0, N-1].$$

(b) The proof of part (b) is completely analogous to that of part (a) and, consequently, is left as an exercise for the reader.   Q.E.D.

We note that the sequence presented in Table 8.2 is real; its transform is given in Table 8.3. Clearly, the results of part (a) of Theorem 8.13 hold, that is,

$$F(1) = 0.08832 - i0.03293 = F^*(15),$$
$$F(2) = 0.06871 - i0.02986 = F^*(14),$$
$$F(3) = 0.05861 - i0.02422 = F^*(13),$$

etc.

We next consider the discrete Fourier transform of an imaginary sequence.

Theorem 8.14. Assume that $\{F(j)\}$ is the discrete Fourier transform of the Nth-order sequence $\{f(k)\}$; then:

(a) If $\{f(k)\}$ is an imaginary sequence, then $\{F(j)\} = -\{F^*(-j)\}$.
(b) If $\{F(j)\}$ is an imaginary sequence, then $\{f(k)\} = -\{f^*(-k)\}$.

Proof.
    (a) By assumption, $\{f(k)\}$ is imaginary, which implies that $\{f^*(k)\} = -\{f(k)\}$. Now we begin with the basic definition of equation (8.5), that is,

$$F(j) = \frac{1}{N} \sum_{k=0}^{N-1} f(k) W_N^{-kj}, \qquad j \in [0, N-1].$$

Taking the complex conjugate of both sides of the above equation yields

$$F^*(j) = \frac{1}{N} \sum_{k=0}^{N-1} f^*(k) W_N^{kj} = -\frac{1}{N} \sum_{k=0}^{N-1} f(k) W_N^{kj}, \qquad j \in [0, N-1].$$

Finally, substitution of $-j$ for j in this equation yields our desired results, that is,

$$F^*(-j) = -\frac{1}{N} \sum_{k=0}^{N-1} f(k) W_N^{-kj} = -F(j), \qquad j \in [0, N-1].$$

    (b) The proof of part (b) is completely analogous to that of part (a) and, consequently, is left as an exercise for the reader.   Q.E.D.

    We conclude this section with a theorem that deals with the transform symmetry of combinations of real and imaginary, as well as odd and even, sequences.

Theorem 8.15. Assume that $\{F(j)\}$ is the discrete Fourier transform of the Nth-order sequence $\{f(k)\}$; then:

(a) If $\{f(k)\}$ is real and even, then $\{F(j)\}$ is real and even.
(b) If $\{f(k)\}$ is real and odd, then $\{F(j)\}$ is imaginary and odd.
(c) If $\{f(k)\}$ is imaginary and even, then $\{F(j)\}$ is imaginary and even.
(d) If $\{f(k)\}$ is imaginary and odd, then $\{F(j)\}$ is real and odd.

Proof.
    (a) By assumption, we have

$$\{f(k)\} \text{ is even, which implies } \{f(k)\} = \{f(-k)\},$$

and

$$\{f(k)\} \text{ is real, which implies } \{f^*(k)\} = \{f(k)\}.$$

Combining the above two results, we have

$$\{f^*(-k)\} = \{f(k)\}.$$

In other words, a sequence is real and even if and only if the above equation is valid. Thus,

$$\{F(j)\} = \mathcal{F}[\{f(k)\}] = \mathcal{F}[\{f^*(-k)\}].$$

However, using Theorem 8.11, we know $\mathcal{F}[\{f^*(-k)\}] = \{F^*(-j)\}$. Thus we obtain our desired results:

$$\{F(j)\} = \{F^*(-j)\},$$

which implies that the sequence $\{F(j)\}$ is real and even.

The proofs of parts (b), (c), and (d) are similar and are left as an exercise.   Q.E.D.

The "pulse" sequence given in Table 8.4 and displayed in Figure 8.6a is real and even. We note that its transform (Table 8.5 and Figure 8.6b) is also real and even as Theorem 8.15(a) states that it must be.

CONVOLUTION OF TWO SEQUENCES

In this section we discuss the convolution summation of two sequences. By definition, the convolution product of two Nth-order sequences $\{f(k)\}$ and $\{g(k)\}$, denoted as $\{f(k)\} * \{g(k)\}$, is given as

$$\{f(k)\} * \{g(k)\} = \sum_{i=0}^{N-1} f(i)g(k-i), \qquad k \in [0, N-1]. \tag{8.11}$$

We note that in this definition the sequence $\{f(k)\}$ ranges over its domain $[0, N-1]$. On the other hand, the domain of of the sequence $\{g(k)\}$ must be extended to include values outside of $[0, N-1]$ since the above summation uses the expression $g(k-i)$. Consequently, we use Theorems 8.2 and 8.3 and consider both sequences $\{f(k)\}$ and $\{g(k)\}$ to be periodic.

For a periodic sequence $\{f(k)\}$, let us consider the following summation:

$$S = \sum_{k=0}^{N-1} f(k). \tag{8.12}$$

If we shift the interval of the summation to the right by an (integer) amount m and sum the terms from $k = m$ to $k = N + m - 1$, then we obtain the same result. This is demonstrated as follows:

$$S = \sum_{k=m}^{N+m-1} f(k) = \sum_{k=m}^{N-1} f(k) + \sum_{k=N}^{N+m-1} f(k).$$

However, since $f(k + N) = f(k)$, the second summation on the right-hand side can be written as

$$\sum_{k=N}^{N+m-1} f(k) = \sum_{k=0}^{m-1} f(k)$$

and thus we obtain

$$\sum_{k=m}^{N+m-1} f(k) = \sum_{k=0}^{m-1} f(k) + \sum_{k=m}^{N-1} f(k) = \sum_{k=0}^{N-1} f(k). \qquad (8.13)$$

The same results can be obtained if we shift the interval to the left by an (integer) amount m and then sum the terms from $k = -m$ to $k = N - m - 1$, that is,

$$\sum_{k=-m}^{N-m-1} f(k) = \sum_{k=0}^{N-1} f(k) = S. \qquad (8.14)$$

With the above results in hand we are able to demonstrate that the convolution product of two sequences is commutative, that is,

$$\{f(k)\} * \{g(k)\} = \{g(k)\} * \{f(k)\}.$$

This is easily demonstrated as follows:

$$\{f(k)\} * \{g(k)\} = \sum_{i=0}^{N-1} f(i)g(k - i), \qquad k \in [0, N - 1].$$

If we now let $n = k - i$ (which implies that $i = k - n$) in the above equation, we obtain

$$\{f(k)\} * \{g(k)\} = \sum_{n=k}^{N-1+k} f(k - n)g(n).$$

Now, using equation (8.13), we obtain

$$\{f(k)\} * \{g(k)\} = \sum_{n=0}^{N-1} g(n)f(k - n) = \{g(k)\} * \{f(k)\}.$$

We now consider the discrete Fourier transform of the convolution of two Nth-order sequences.

Theorem 8.16 (Convolution). If both Nth-order sequences $\{f(k)\}$ and $\{g(k)\}$ have discrete Fourier transforms given by $\{F(j)\}$ and $\{G(j)\}$, respectively, then the discrete Fourier transform of the convolution product of $\{f(k)\}$ and $\{g(k)\}$ is given as the simple product of the individual transforms, that is,

$$\mathcal{F}[\{f(k)\} * \{g(k)\}] = N\{F(j)\}\{G(j)\}.$$

Proof. We begin by considering the product of the transforms $\{F(j)\}$ and $\{G(j)\}$, that is,

$$\{F(j)\}\{G(j)\} = \left[\frac{1}{N}\sum_{n=0}^{N-1}f(n)W_N^{-nj}\right]\left[\frac{1}{N}\sum_{k=0}^{N-1}g(k)W_N^{-kj}\right],$$

$$\{F(j)\}\{G(j)\} = \frac{1}{N^2}\sum_{n=0}^{N-1}\left[\sum_{k=0}^{N-1}f(n)g(k)\right]W_N^{-(k+n)j}.$$

If we now make the substitution $i = k + n$ (which implies that $k = i - n$), we find

$$\{F(j)\}\{G(j)\} = \frac{1}{N^2}\sum_{n=0}^{N-1}\left[\sum_{i=n}^{N+n-1}f(n)g(i-n)\right]W_N^{-ij}.$$

Now, interchanging the summations and using equation (8.13) on the periodic sequence $f(n)g(i-n)$, we obtain

$$\{F(j)\}\{G(j)\} = \frac{1}{N}\sum_{i=0}^{N-1}\left[\sum_{n=0}^{N-1}\frac{f(n)g(i-n)}{N}\right]W_N^{-ij},$$

or

$$\{F(j)\}\{G(j)\} = \frac{1}{N}\sum_{i=0}^{N-1}\left[\frac{\{f(i)\}*\{g(i)\}}{N}\right]W_N^{-ij}.$$

Thus,

$$\{F(j)\}\{G(j)\} = \mathcal{F}\left[\frac{\{f(k)\}*\{g(k)\}}{N}\right] = \left(\frac{1}{N}\right)\mathcal{F}[\{f(k)\}*\{g(k)\}].\quad \text{Q.E.D.}$$

The dual of this theorem is the product theorem, which tells us that the discrete Fourier transform of the simple product of two sequences is the convolution product of the individual transforms. This is stated formally in the following theorem.

Theorem 8.17 (Product). If both Nth-order sequences $\{f(k)\}$ and $\{g(k)\}$ have discrete Fourier transforms given by $\{F(j)\}$ and $\{G(j)\}$, respectively, then the discrete Fourier transform of the product of $\{f(k)\}$ and $\{g(k)\}$ is the convolution product of the respective transforms, that is,

$$\mathcal{F}[\{f(k)\}\{g(k)\}] = \{F(j)\}*\{G(j)\}.$$

Proof. We begin by considering the product of $\{f(k)\}$ and $\{g(k)\}$ in terms of the inverse discrete Fourier transform expression of equation (8.6), that is,

$$\{f(k)\}\{g(k)\} = \left[\sum_{n=0}^{N-1} F(n)W_N^{kn}\right]\left[\sum_{j=0}^{N-1} G(j)W_N^{kj}\right],$$

$$\{f(k)\}\{g(k)\} = \sum_{n=0}^{N-1}\sum_{j=0}^{N-1} F(n)G(j)W_N^{(n+j)k}.$$

If we now let $i = n + j$ (which implies $j = i - n$), then we obtain

$$\{f(k)\}\{g(k)\} = \sum_{n=0}^{N-1}\sum_{i=n}^{N+n-1} F(n)G(i-n)W_N^{ik}.$$

Now, interchanging the order of summation and using equation (8.13) on the periodic sequence $\{F(n)G(i-n)\}$, we obtain

$$\{f(k)\}\{g(k)\} = \sum_{i=0}^{N-1}\left[\sum_{n=0}^{N-1} F(n)G(i-n)\right]W_N^{ik},$$

$$\{f(k)\}\{g(k)\} = \sum_{n=0}^{N-1}[\{F(i)\} * \{G(i)\}]W_N^{ik} = \mathcal{F}^{-1}[\{F(i)\} * \{G(i)\}].$$

Taking the discrete Fourier transform of both sides of the above equation yields our desired results, that is,

$$\mathcal{F}[\{f(k)\}\{g(k)\}] = \{F(j)\} * \{G(j)\}. \qquad \text{Q.E.D.}$$

SIMULTANEOUS CALCULATION OF REAL TRANSFORMS

In this section we demonstrate how Theorems 8.4 and 8.13 may be combined to obtain a very useful computational technique. This technique permits the simultaneous calculation of the discrete Fourier transform of two real sequences. This method is of significant practical interest because many times two real sequences are transformed to perform a frequency domain convolution or deconvolution and most fast Fourier transform codes (see next section) assume that a complex function is to be transformed (the significance of this last comment will become apparent as we proceed). We begin by considering two sequences $\{f_1(k)\}$ and $\{f_2(k)\}$ with discrete Fourier transforms $\{F_1(j)\}$ and $\{F_2(j)\}$, respectively. We first combine these two real sequences to obtain the complex sequence $\{f(k)\}$:

$$\{f(k)\} = \{f_1(k)\} + i\{f_2(k)\}. \qquad (8.15)$$

TABLE 8.8 Example 7

j	Real	Imaginary
0	.13119	.31250
1	.27668	.15543
2	.21959	−.02986
3	.07413	−.03974
4	.05284	.04384
5	.10152	.03859
6	.07323	−.00881
7	.02121	.02070
8	.04590	.06250
9	.07128	.02937
10	.02145	.00881
11	−.00274	.06567
12	.05284	.08116
13	.04309	.00870
14	−.08218	.02986
15	−.10003	.22128

Using Theorem 8.4 (linearity), we have

$$\{F(j)\} = \mathcal{F}[\{f(k)\}] = \{F_1(j)\} + i\{F_2(j)\}. \tag{8.16}$$

Taking the complex conjugate of both sides of this equation, we obtain

$$\{F^*(j)\} = \{F_1^*(j)\} - i\{F_2^*(j)\}. \tag{8.17}$$

We now substitute $N - j$ for j in equation (8.17) and obtain

$$\{F^*(N - j)\} = \{F_1^*(N - j)\} - i\{F_2^*(N - j)\}. \tag{8.18}$$

However, since $\{f_1(k)\}$ and $\{f_2(k)\}$ are real sequences, we know from Theorem 8.13 that

$$\{F_1(j)\} = \{F_1^*(-j)\} = \{F_1^*(N - j)\},$$
$$\{F_2(j)\} = \{F_2^*(-j)\} = \{F_2^*(N - j)\}.$$

Substituting these expressions into equation (8.18) yields

$$\{F^*(N - j)\} = \{F_1(j)\} - i\{F_2(j)\}. \tag{8.19}$$

Addition and subtraction of equations (8.16) and (8.19) produces our desired result:

$$\{F_1(j)\} = \frac{[\{F(j)\} + \{F^*(N - j)\}]}{2},$$

$$\{F_2(j)\} = \frac{[\{F(j)\} - \{F^*(N - j)\}]}{2i}. \tag{8.20}$$

Example 7. Let us use this technique to determine the Fourier transform of sequences given in Tables 8.2 and 8.6. To do so, we first form the new sequence

$$\{f(k)\} = \{f_1(k)\} + i\{f_2(k)\}, \qquad k \in [0, N-1],$$

where $\{f_1(k)\}$ denotes the sequence given in Table 8.2; similarly, $\{f_2(k)\}$ denotes the sequence presented in Table 8.6. Using equation (8.1), we calculate the discrete Fourier transform of this sequence; the results are presented in Table 8.8. Now using equations (8.20) we see that

$$F_1(0) = [F(0) + F^*(16)]/2$$
$$= [(0.13119 + i0.31250) + (0.1339 - i0.31250)]/2$$
$$= 0.13119,$$
$$F_1(1) = [F(1) + F^*(15)]/2$$
$$= [(0.27668 + i0.15543) + (-0.1003 - i0.22128)]/2$$
$$= 0.8832 - i0.03293,$$
$$F_1(2) = [F(2) + F^*(14)]/2$$
$$= [(0.21959 - i0.02986) + (-0.08218 - i0.02986)]/2$$
$$= 0.06871 - i0.02986,$$

etc.

Comparing these results to those presented in Table 8.3, we see that they agree with our previous calculation of $\{F_1(j)\}$.

Continuing, we obtain

$$F_2(0) = [F(0) - F^*(16)]/2i$$
$$= [(0.13119 + i0.31250) - (0.1339 - i0.31250)]/2i$$
$$= 0.31250,$$
$$F_2(1) = [F(1) - F^*(15)]/2i$$
$$= [(0.27668 + i0.15543) - (-0.1003 - i0.22128)]/2i$$
$$= 0.18835 - i0.18835,$$
$$F_2(2) = [F(2) - F^*(14)]/2i$$
$$= [(0.21959 - i0.02986) - (-0.08218 - i0.02986)]/2i$$
$$= 0.00000 - i0.15089,$$

etc.

Again, we see that these results agree with those previously calculated and presented in Table 8.7.

THE FAST FOURIER TRANSFORM

In 1965, J. W. Tukey and J. W. Cooley published an algorithm that, under certain conditions, tremendously reduces the number of computations required to compute the discrete Fourier transform of a sequence. This algorithm has come to be called the *fast Fourier transform* and is considered one of the most significant contributions to numerical analysis of this century. Basically, the fast Fourier transform is a clever computational technique of sequentially combining progressively larger weighted sums of data samples so as to produce the discrete Fourier transform. These comments become more apparent as we proceed.

We begin by assuming that we have an Nth-order sequence $\{f(k)\}$ with discrete Fourier transform $\{F(j)\}$. Furthermore, let us assume that N is an even integer, and thus we are able to form two new subsequences

$$\{f_1(k)\} = \{f(2k)\},$$
$$\{f_2(k)\} = \{f(2k+1)\}, \qquad k = 0, 1, \ldots, M-1, \quad \text{where} \quad M = N/2. \tag{8.21}$$

We note here that

$$\{f_1(k+M)\} = f(2(k+M)) = f(2k+N) = f(2k) = \{f_1(k)\},$$
$$\{f_2(k+M)\} = f(2(k+M)+1) = f(2k+N+1) = f(2k+1) = \{f_2(k)\},$$

and therefore we see that both $\{f_1(k)\}$ and $\{f_2(k)\}$ are periodic sequences with periodicity M.

As an example, consider the 8th-order sequence

$$\{f(k)\} = \{f(0), f(1), f(2), f(3), f(4), f(5), f(6), f(7)\}.$$

In this case we have

$$\{f_1(k)\} = \{f(0), f(2), f(4), f(6)\},$$
$$\{f_2(k)\} = \{f(1), f(3), f(5), f(7)\}.$$

Since $\{f_1(k)\}$ and $\{f_2(k)\}$ are Mth-order sequences we can use equation (8.5) to determine their discrete Fourier transform:

$$F_1(j) = \frac{1}{M} \sum_{k=0}^{M-1} f_1(k) W_M^{-kj}, \qquad j \in [0, N-1],$$

$$F_2(j) = \frac{1}{M} \sum_{k=0}^{M-1} f_2(k) W_M^{-kj}, \qquad j \in [0, N-1]. \tag{8.22}$$

We note that by Theorems 8.2 and 8.3 both $\{F_1(j)\}$ and $\{F_2(j)\}$ may be considered periodic with periodicity M. Now let us consider the discrete Fourier transform of the Nth-order sequence $\{f(k)\}$:

$$F(j) = \frac{1}{N} \sum_{k=0}^{N-1} f(k) W_N^{-kj}. \tag{8.23}$$

Splitting the summation, we can rewrite the preceding equation as

$$F(j) = \frac{1}{N} \sum_{k=0}^{M-1} f(2k) W_N^{-2kj} + \frac{1}{N} \sum_{k=0}^{M-1} f(2k+1) W_N^{-(2k+1)j}.$$

However, we note that

$$W_N^{-2kj} = e^{-2\pi i kj/(N/2)} = W_M^{-kj},$$

$$W_N^{-(2k+1)j} = e^{-2\pi i(2k+1)j/N} = W_M^{-kj} W_N^{-j}.$$

Therefore, the previous equation becomes

$$F(j) = \frac{1}{N} \sum_{k=0}^{M-1} f_1(k) W_M^{-kj} + \frac{W_N^{-j}}{N} \sum_{k=0}^{M-1} f_2(k) W_M^{-kj}, \qquad j \in [0, N-1].$$

Comparison to equations (8.22) yields

$$F(j) = \frac{F_1(j)}{2} + \frac{W_N^{-j} F_2(j)}{2}, \qquad j \in [0, N-1]. \tag{8.24}$$

Because $\{F_1(j)\}$ and $\{F_2(j)\}$ are periodic with (with period M) we have

$$F(j) = \tfrac{1}{2}[F_1(j) + F_2(j) W_N^{-j}], \qquad j \in [0, M-1],$$
$$F(j+M) = \tfrac{1}{2}[F_1(j) - F_2(j) W_N^{-j}], \qquad j \in [0, M-1]. \tag{8.25}$$

As we have noted, to calculate the discrete Fourier transform of $\{f(k)\}$ requires N^2 complex additions and multiplications, whereas to calculate the discrete Fourier transform of $\{f_1(k)\}$ or $\{f_2(k)\}$ requires only M^2 or $N^2/4$ complex operations. When we use equation (8.25) to obtain $\{F(j)\}$ from $\{F_1(j)\}$ and $\{F_2(j)\}$, we require $N + 2(N^2/4)$ complex operations. In other words, we first require $2(N^2/4)$ operations to calculate the two Fourier transforms $\{F_1(j)\}$ and $\{F_2(j)\}$, and then we require the N additional operations prescribed by equation (8.25). Thus, we have reduced the number of operations from N^2 to $N + N^2/2$. For the smallest value of N (i.e., $N = 4$), this results in a factor of 0.75. When N is large, this factor approaches 1/2.

Before proceeding, we make note of the important fact that when $N = 2$ we divide the second-order sequence $\{f(k)\} = \{f(0), f(1)\}$ into two first-order sequences $f_1(0) = f(0)$ and $f_2(0) = f(1)$. However, since a first-order sequence is its own transform [i.e., $F_1(0) = f_1(0)$ and $F_2(0) = f_2(0)$], we do not require any complex multiplications or additions to obtain these transforms. Therefore, using equation (8.25) for this case would require only $N = 2$ operations to obtain $F(j)$ rather than $N = 4$ if we applied equation (8.5).

Next, suppose that N is divisible by 4 or that $M = N/2$ is divisible by 2. Then the subsequences $\{f_1(k)\}$ and $\{f_2(k)\}$ can be further subdivided into four $M/2$-order sequences as per equation (8.21):

$$g_1(k) = f_1(2k),$$

$$g_2(k) = f_1(2k + 1),$$

$$h_1(k) = f_2(2k),$$

$$h_2(k) = f_2(2k + 1),$$

for $k = 0,1,\ldots,M/2-1$. For example, the 8th-order sequence $\{f(k)\}$ is reduced as follows:

$$\{g_1(k)\} = \{f(0), f(4)\},$$

$$\{g_2(k)\} = \{f(2), f(6)\},$$

$$\{h_1(k)\} = \{f(1), f(5)\},$$

$$\{h_2(k)\} = \{f(3), f(7)\}.$$

It should be rather obvious that we can also use equation (8.25) to obtain the discrete Fourier transforms of $\{F_1(j)\}$ and $\{F_2(j)\}$ with only $M + M^2/2$ complex operations and then use these results to obtain $\{F(j)\}$. A little thought reveals that this requires $N + 2(M + M^2/2) = 2N + N^2/4$ operations.

Thus, when we subdivide a sequence twice ($N > 4$ and N divisible by 4) we reduce the number of operations from N^2 to $2N + N^2/4$. The $2N$ term is the result of applying equation (8.25) (twice), whereas the $N^2/4$ term is the result of transforming the four reduced sequences. For the case when $N = 4$ we note that we completely reduce the sequence to four first-order sequences that are their own transforms, and therefore we do not need the additional $N^2/4$ transform operations. The formula then becomes $2N$. The smallest value of N that does not result in complete reduction of the sequence is $N = 8$. For this case we have a reduction factor of 1/2, whereas for large N the factor approaches 1/4. Continuing in this way, we can show that if N is divisible by 2^p (where p is a positive integer), then the number of operations required to compute the discrete Fourier transform of the Nth-order sequence $\{f(k)\}$ by repeated subdivision is

$$pN + N^2/2^p.$$

Again for complete reduction (i.e., $N = 2^p$) the $N^2/2^p$ term is not required, and we obtain pN for the number of operations required. This results in a reduction factor of

$$pN/N^2 = p/N = \log_2(N)/N. \tag{8.26}$$

The essence of the Cooley–Tukey algorithm is to choose sequences with $N = 2^p$ and go to complete reduction. Although most sequences do not have such a convenient number of terms, we can always artificially add zeros to the end (or begining) of the sequence to reach such a value. This extra number of terms in the sequence is more than compensated for by the tremendous savings afforded by using the Cooley–Tukey algorithm. As an example, for values of $N = 512$, 1024, and 2048 we obtain reduction factors of approximately 57, 102, and 186, respectively. While, at first glance, these values may not seem significant, a 2048-point discrete Fourier transform that may take a full hour to perform on a small PC would only require 19 seconds using the fast Fourier transform algorithm.

SUMMARY

In this chapter we presented the discrete Fourier transform, which is an operation that maps an Nth-order sequence to another Nth-order sequence. We then presented several theorems that described various properties of this transform.

It is very important to note that the sequences involved need not be periodic for the discrete Fourier transform to be defined. However, many of the properties that we presented require that we extend the domain of the sequences (e.g., the shifting theorems), and to do this we used a periodic extension. Consequently, to apply many of the theorems presented in this chapter we do actually assume that the sequences are periodic.

PROBLEMS

1   Several of the following problems require you to calculate the discrete Fourier transform of an Nth-order sequence. To accomplish this you will need to write an algorithm, or computer program, that will calculate the real and imaginary portions of the transform as given per equation (8.3).

2   Calculate the discrete Fourier transform of the 16th-order sequence $\{f(k)\}$ whose terms are given as $f(k) = \exp[-k/2]$.

3   Calculate the discrete Fourier transform of the 16th-order sequence $\{f(k)\}$ whose terms are given as $f(k) = \cos(k)$.

4   Calculate the discrete Fourier transform of the 16th-order sequence $\{f(k)\}$ whose terms are given as $f(k) = \mathrm{sinc}(2\pi k/8)$.

5   Calculate the discrete Fourier transform of the 16th-order sequence $\{f(k)\}$ whose terms are given as $f(k) = k^2$.

6   Calculate the discrete Fourier transform of the 16th-order sequence $\{f(k)\}$ whose terms are given as $f(k) = \exp[-(k-8)^2]$.

7 Use equation (8.3) to find the discrete Fourier transform of $\exp[-(k-3)/2]$. Now use the results of problem 2 and the first shifting theorem (Theorem 8.5) to obtain the answer.

8 Prove Theorem 8.12, part (b).

9 Prove Theorem 8.13, part (b).

10 Prove Theorem 8.14, part (b).

11 Prove Theorem 8.15, parts (b), (c), and (d).

12 Consider the sequence $\{\cos(2\pi kn/N)\}$. What is the first forward difference of this sequence when $n = 4$ and $N = 16$? What is the discrete Fourier transform of the sequence and its first forward difference?

13 Consider the sequences $\{f(k)\} = \{k\}$ and $\{g(k)\} = \{1/(k+1)\}$. Using equation (8.11), determine the convolution of $\{f(k)\}$ and $\{g(k)\}$ for $N = 16$.

14 Use the convolution theorem (Theorem 8.16) to calculate the convolution product of problem 13.

15 Show that the delta sequence is the unit element under the convolution product of equation (8.11).

16 Show that if N is an even integer and $\{f(k)\}$ is an odd sequence, then $f(N/2) = 0$.

17 Show that the convolution summation of equation (8.11) is associative, that is,

$$\{f(k)\} * [\{g(k)\} * \{h(k)\}] = [\{f(k)\} * \{g(k)\}] * \{h(k)\}.$$

18 Show that the convolution product of equation (8.11) is distributive with respect to addition, that is,

$$\{f(k)\} * [\{g(k)\} + \{h(k)\}] = \{f(k)\} * \{g(k)\} + \{f(k)\} * \{h(k)\}.$$

BIBLIOGRAPHY

Cooley, J. W., and J. W. Tukey, "An Algorithm for the Machine Calculation of Complex Fourier Series," *Mathematical Computations*, **19**, April 1965.

Cooley, J. W., P. A. W. Lewis, and P. D. Welsh, "The Finite Fourier Transform," *IEEE Transactions on Audio-Electroacoustics*, **AU-17**, No. 2, June 1969.

Freeman, H., *Discrete Time Systems, An Introduction to the Theory*, John Wiley & Sons, New York, 1965.

G-AE Subcommittee on Measurement Concepts, "What is the Fast Fourier Transform," *IEEE Transactions on Audio and Electroacoustics*, **AU- 15**, No. 2, June 1967.

Singleton, R. C., "A Short Bibliography on the Fast Fourier Transform," *IEEE Transactions on Audio and Electroacoustics*, **AU-17**, No. 2, June 1966.

Weaver, H. J., *Applications of Discrete and Continuous Fourier Analysis*, John Wiley & Sons, New York, 1983.

9

SAMPLING THEORY

Up to this point in the text we have discussed the Fourier series, the Fourier transform, and the discrete Fourier transform. The Fourier series may be considered a transformation that maps a periodic function to a sequence of discrete coefficients. The Fourier transform is a mapping that takes a function (or more generally a distribution) $f(t)$ to another distribution $F(w)$. Finally, in the last chapter we presented the discrete Fourier transform that maps a sequence $\{f(k)\}$ to its transform sequence $\{F(j)\}$.

Both the Fourier series and the Fourier transform require the analytical determination of an integral equation. Depending upon the complexity of the function involved, these integrals can be rather difficult, if not impossible, to evaluate. Furthermore, in the real world we are interested, more often than not, in obtaining the Fourier transform (or series) of signals that are obtained from some physical phenomenon or experiment, and these signals cannot usually be described by any analytical formula. For example, we may have reason to determine the frequency content of a signal that is obtained from an accelerometer that is mounted on the axle of a truck that is driven over a bumpy country road.

If we compare the basic definition and properties (e.g., the shifting theorems) for the Fourier series, the Fourier transform, and the discrete Fourier transform, we note rather obvious similarities among them. In this chapter we take advantage of these similarities and demonstrate how the relatively simple discrete Fourier transform operation can be used to obtain both the Fourier transform and Fourier series of a signal or function. The basic approach that we use is to convert the function to a sequence (by sampling), calculate its discrete Fourier transform, and then relate this transformed sequence to the Fourier transform (or Fourier series) of the original function.

SAMPLING A FUNCTION

When we sample a function $f(t)$, with sampling rate Δt, we convert the function to a sequence $\{f(k)\}$ whose terms are values of the function at the discrete locations $k\Delta t$. In other words, the terms of the sequence, denoted as $f(k)$, are given as $f(t)$ evaluated at $t = k\Delta t$. As an example, consider the function $f(t) = \exp[-t^2]$ defined over the domain $[-3,3]$ and shown graphically in Figure 9.1a. If we sample this function with sampling rate $\Delta t = 0.15$ over this domain, then we obtain a 40th-order sequence $\{f(k)\}$ that is shown in Figure 9.1b.

As we learned in Chapter 7, when we multiply a function $f(t)$ by a comb function with spacing Δt we obtain a sequence $\{f(k)\}$. In other words, multiplying a function by a comb function is the mathematical equivalent of sampling, that is,

$$\{f(k)\} = f(t)\mathrm{comb}_{\Delta t}(t) = \sum_{k=-\infty}^{\infty} f(t)\delta(t - k\Delta t). \tag{9.1}$$

The first question that should come to mind is: What sampling rate Δt is required to properly sample the function $f(t)$? To answer this question we must first decide what constitutes a proper sampling of a function. Let us agree that a function is properly sampled if that function can be recovered exactly from its sampled sequence—that is to say, if we can somehow interpolate between the sequence terms $f(k)$ to return the function $f(t)$.

We now proceed to show that if the function $f(t)$ is band limited with bandwidth Ω and the sampling rate is chosen such that $\Delta t = 1/2\Omega$, then $f(t)$ can be recovered uniquely and exactly from the sampled sequence $\{f(k)\}$.

We first note that a function is called "band limited with bandwidth 2Ω" if its Fourier transform $F(w)$ has bounded support over the interval $[-\Omega,\Omega]$. In other words, $F(w) = 0$ (in the sense of the L-1 norm) for values of w outside the interval $[-\Omega,\Omega]$.

We begin by taking the Fourier transform of both sides of equation (9.1) and using the product theorem (Theorem 9.13), along with the fact that the Fourier transform of $\mathrm{comb}_{\Delta t}(t)$ is given as $\Delta w\,\mathrm{comb}_{\Delta w}(w)$, where $\Delta w = 1/\Delta t$. This move results in

$$\mathcal{F}[f(t)\mathrm{comb}_{\Delta t}(t)] = (1/\Delta t)F(w)*\mathrm{comb}_{\Delta w}(w). \tag{9.2}$$

Now recall the fact that when we convolve a function $F(w)$ with a comb function of spacing Δw we copy the function $F(w)$ to locations $w = j\Delta w$. The transform of equation (9.2) is shown graphically in Figure 9.2. (Note that the amplitude of each "copy" is reduced by a factor of Δt.) We now wish to recover $F(w)$ (or, equivalently, $f(t)$, using the inverse Fourier transform) from

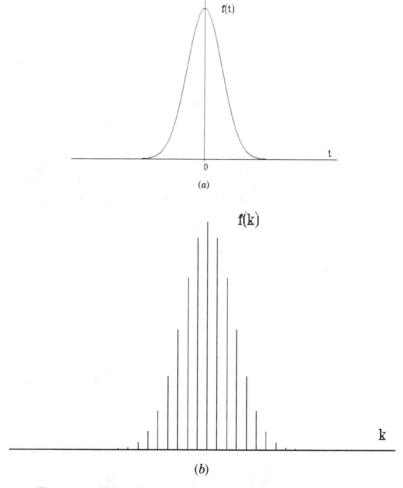

Figure 9.1 (a) The function $f(t)$ and (b) its sampled sequence $\{f(k)\}$.

the function shown in Figure 9.2. There are several such ways to accomplish this recovery; however, the simplest method is to multiply the function by a pulse function of amplitude Δt and half-width Ω to obtain $F(w)$ and then inverse Fourier transform the results. This multiplication recovery scheme is illustrated in Figure 9.3. Mathematically, that is to say,

$$F(w) = [\Delta w F(w) * \text{comb}_{\Delta w}(w)] p_\Omega(w) \Delta t,$$

or

$$f(t) = \mathcal{F}^{-1}[(\Delta w F(w) * \text{comb}_{\Delta w}(w)] * \mathcal{F}^{-1}[p_\Omega(w)] \Delta t,$$

where $\Delta w = 1/\Delta t$. However, it can easily be demonstrated that

$$\mathcal{F}^{-1}[P_\Omega(w)] \Delta t = \text{sinc}(2\pi\Omega t).$$

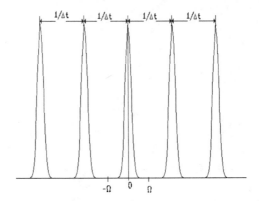

Figure 9.2 Fourier transform of sampled sequence as per equation (9.2).

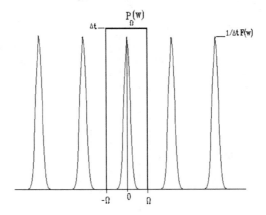

Figure 9.3 Sampling theorem recovery scheme in the frequency domain.

Thus,

$$f(t) = [f(t)\,\text{comb}_{\Delta t}(t)] * \text{sinc}(2\pi\Omega t),$$

or

$$f(t) = \sum_{k=-\infty}^{\infty} f(t)\delta(t - k\Delta t) * \text{sinc}(2\pi\Omega t),$$

$$f(t) = \sum_{k=-\infty}^{\infty} f(k\Delta t)\,\text{sinc}(2\pi\Omega[t - k\Delta t]). \tag{9.3}$$

The results of the above equation are known as the Whittaker–Shannon sampling theorem, which tells us that when Δt is chosen such that $\Delta t = 1/2\Omega$, the function $f(t)$ can be recovered uniquely and exactly from the values of the sampled sequence $\{f(k)\}$ by using the interpolation formula

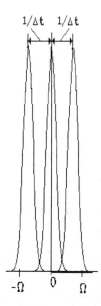

$1/\Delta t$ $1/\Delta t$

$-\Omega$ 0 Ω

Figure 9.4   Aliasing demonstrated in the frequency domain.

given in equation (9.3). This sampling rate has come to be called the *Nyquist rate*, in honor of the early frequency domain pioneer Harry Nyquist (1889–1976).

ALIASING

In Figure 9.2 we see the significance of choosing the sampling rate $\Delta t \leq 1/2\Omega$. If $\Delta t > 1/2\Omega$ (or, equivalently, $1/\Delta t < 2\Omega$), then the repeated transforms would not be separated but, instead, would overlap as shown in Figure 9.4. In this case we cannot use the pulse function masking technique to recover $F(w)$ uniquely because the higher components of $F(w - \Delta w)$ and $F(w + \Delta w)$ will overlap, or spill into the basic transform $F(w)$ centered at $w = 0$. This condition is known as *aliasing*. In the time domain, aliasing shows up as a sampling rate that is too coarse, such as the one illustrated in Figure 9.5.

One technique that is commonly used to avoid alising is to first pass the signal through a low-pass (analog) anti-aliasing filter before it is sampled. The filter bandwidth is tied directly to the sampling rate and is set such that it removes all frequencies above $\Omega = 1/2\Delta t$. In this way, the sampling theorem Nyquist condition $\Delta t \leq 1/2\Omega$ is always satisfied. In other words, the sampling rate dictates the bandwidth. Obviously, when this type of filtering technique is used we run the risk of not sampling the correct function. In other words, if the function does indeed contain frequencies above Ω, then they will be

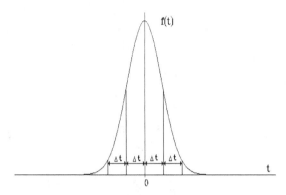

Figure 9.5   Aliasing demonstrated in the time domain.

removed and we will be sampling a function that could be substantially differ-ent from the original one. From a practical point of view, the sampling rate (and consequently the anti-aliasing filter bandwidth) are sequentially adjusted finer and finer until the sampled function function does not change with any further adjustment in the sampling rate.

We also point out here that the Nyquist rate is the theoretical limit that should be used as a guide when sampling a function and should not be taken as a sacred or absolute value. In other words, by definition, the band limit of a function is that frequency value Ω such that the function is zero outside of the interval $[-\Omega, \Omega]$. This need not be a fixed or set value inasmuch as if $\Omega_1 > \Omega$ and Ω is a bandlimit, then so to is Ω_1 because the function will also be zero outside the interval $[-\Omega_1, \Omega_1]$. In actual practice, in the real world a sampling rate of five times the Nyquist rate is usually recommended so as to avoid any problems caused by noise or other signal degradation. We call this sampling rate ($\Delta t = 1/10\Omega$) the *real-world sampling rate*.

COMPUTER CALCULATION OF THE FOURIER TRANSFORM

In this section we describe how the Fourier transform of a function $f(t)$ can be obtained from the discrete Fourier transform of the sequence $\{f(k)\}$ formed from the samples of the original function. We begin by assuming that the function $f(t)$ is of bounded support as well as being band limited. In other words, assume that $f(t)$ has bounded support over the time domain interval $[0, T]$ and is also band limited over the frequency domain interval $[-\Omega, \Omega]$. Next, let us sample the function with the Nyquist rate $\Delta t = 1/2\Omega$ and choose the number of samples such that

$$N\Delta t \geq 2T.$$

Thus, using the sampling theorem, we are able to recover the function $f(t)$ from its sampled sequence $\{f(k)\}$ using the following interpolation formula:

$$f(t) = \sum_{k=0}^{N-1} f(k)\,\mathrm{sinc}(2\pi\Omega(t - k\Delta t)).$$

In the above equation we have the function $f(t)$ described in terms of its sampled sequence $\{f(k)\}$. We now Fourier transform both sides of the equation (with respect to t), that is,

$$F(w) = \sum_{k=0}^{N-1} f(k)\mathcal{F}[\mathrm{sinc}(2\pi\Omega(t - k\Delta t))].$$

Using the scale change theorem (Theorem 7.2) with scale factor $2\pi\Omega$, the first shifting theorem (Theorem 7.3) with shift factor $k\Delta t$, and the fact that the Fourier transform of $\mathrm{sinc}(t)$ is given as $\pi p_{1/2\Omega}(w)$, we obtain

$$F(w) = \frac{1}{2\Omega} \sum_{k=0}^{N-1} f(k)e^{-2\pi i w k\Delta t} p_{1/2\Omega}(w).$$

However, we are only interested in $F(w)$ over the bandwidth $[-\Omega,\Omega]$, for which $p_{1/2\Omega}(w) = 1$. Thus, the above equation becomes

$$F(w) = \frac{1}{2\Omega} \sum_{k=0}^{N-1} f(k)e^{-2\pi i w k\Delta t}, \tag{9.4}$$

or

$$F(w) = \frac{1}{2\Omega} \sum_{k=0}^{N-1} f(k)[\cos(2\pi w k\Delta t) - i\sin(2\pi w k\Delta t)]. \tag{9.5}$$

The above two equations give us formulas from which we can obtain the Fourier transform of the function $f(t)$ from the sampled sequence $\{f(k)\}$. Next let us sample $F(w)$ (as given by equation (9.4)) with sampling rate $\Delta w = 2\Omega/N$. We first note that with this sampling rate we can uniquely recover $F(w)$ from its sampled sequence $\{F(j\Delta w)\}$ because

$$\Delta w = 1/(N\Delta t) = 1/2T,$$

which is equal to the bandwidth of $F(w)$ inasmuch as

$$\mathcal{F}[F(w)] = \mathcal{F}[\mathcal{F}[f(t)]] = f(-t).$$

In other words, if we consider $f(-t)$ to be the Fourier transform of $F(w)$, then the bandwidth of $F(w)$ is $2T$ because $f(t)$ is zero outside the interval $[0,T]$.

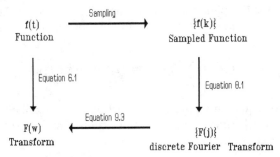

Figure 9.6 Schematic representation of method of using the discrete Fourier transform to obtain the Fourier transform of a function.

Now when we sample equation (9.4) with this sampling rate, we obtain

$$F(j) = \frac{1}{2\Omega} \sum_{k=0}^{N-1} f(k)e^{-2\pi ijk\Delta w \Delta t}.$$

However,

$$\Delta w \Delta t = (2\Omega/N)(1/2\Omega) = 1/N$$

and

$$1/2\Omega = (\Delta t N)(1/N).$$

When these results are substituted into the above equation we obtain

$$F(j) = N\Delta t \left(\frac{1}{N}\right) \sum_{k=0}^{N-1} f(k)e^{-2\pi ikj/N} = N\Delta t \mathcal{F}\{[f(k)]\}.$$

In words, the above equation tells us that the samples $F(j)$ of the Fourier transform $F(w)$ (using sampling rate $\Delta w = 2\Omega/N$) are the same as the discrete Fourier transform of the sampled sequence $\{f(k)\}$ multiplied by $N\Delta t$.

This previous demonstration provides us with a method of obtaining the Fourier transform of a general signal using a digital computer. That is to say, we first sample the function $f(t)$ with the sampling rate $\Delta t = 1/2\Omega$, choosing the number of samples N such that $N\Delta t = 2T$. Next we perform the discrete Fourier transform operation on the sampled sequence $\{f(k)\}$ and multiply the resulting sequence by $N\Delta t$ to obtain the sequence $\{F(j)\}$. The terms of this new sequence are equal to the samples of the Fourier transform $F(w)$ of the original function (sampled with sampling rate $\Delta w = 2\Omega/N = 1/N\Delta t$. Finally, $F(w)$ is recovered from $\{F(j)\}$ by using the sampling theorem interpolation formula. This procedure is shown schematically in Figure 9.6.

To actually use the technique just described, a few facts must be pointed out. Our first assumption was that the function $f(t)$ had bounded support over the interval $[0,T]$. In other words, the function was zero for $t < 0$. This is not always the case; for example, consider either the pulse or Gaussian function.

When this is not the case we simply form a new function $g(t)$ by shifting $f(t)$ to the right by some amout A so that $g(t)$ is of bounded support on the interval $[0, T + A]$. The technique can now be used to obtain the transform of $g(t)$, that is, $G(w) = F(w)\exp[-2\pi i w A]$. To recover our desired result $F(w)$, we rewrite the previous equation as

$$F(w) = G(w)e^{2\pi i w A} = G(w)[\cos(2\pi w A) + i\sin(2\pi w A)].$$

Also, rather than the theoretical Nyquist sampling rate, we should use the real-world or practical rate of $1/10\Omega$ and choose the number of samples N such that $N\Delta t > 10(T + A)$.

Finally, when we obtain the discrete Fourier transform of a sequence, we obtain N terms of the sequence ranging from $j = 0$ to $j = N - 1$. Here we take advantage of Theorem 8.3 to note that $F(-j) = F(N - j)$ and consider the transform from $-N/2$ to $N/2$. The spacing between each sample is $\Delta w = 1/N\Delta t$. In this way the Fourier transform of the original function is obtained.

The overall procedure is summarized in the following six steps.

Assume that $f(t)$ has bounded support over the interval $[-A, T]$. Let us also assume that it is band limited over the frequency domain interval $[-\Omega, \Omega]$. Then, to digitally obtain the Fourier transform of this function, we proceed as follows:

1. If necessary $(A \neq 0)$, form the new function $g(t)$ by shifting $f(t)$ to the right by an amount A:

$$g(t) = f(t - A).$$

2. Sample the function $g(t)$ with sampling rate $\Delta t = 1/10\Omega$ and choose the number of samples N such that

$$N\Delta t > 10(T + A).$$

3. Calculate the discrete Fourier transform of this sampled sequence and multiply the resulting sequence by $N\Delta t$ to obtain the sequence $\{G(j\Delta w)\}$.

4. Use Theorem 8.3 to obtain transform values for the negative indices j that represent values for the negative frequencies $-j\Delta w$ $(\Delta w = 1/N\Delta t)$.

5. Recover $G(w)$ from $\{G(j\Delta w)\}$ as per the sampling theorem interpolation formula (equation (9.4)) (or by simply constructing a smooth curve between the sampled values).

6. If necessary $(A \neq 0)$, recover $F(w)$ from $G(w)$ as per the formula

$$F(w) = G(w)e^{2\pi i A w}.$$

Note: If $A = 0$, then $F(w) = G(w)$.

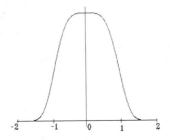

Figure 9.7 Function $f(t)$ to be transformed.

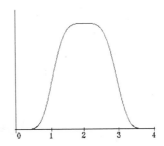

Figure 9.8 The shifted function $g(t) = f(t - 2)$.

As an example, we now obtain the Fourier transform of the function

$$f(t) = \exp[-t^4], \qquad \text{over the interval } [-2, 2].$$

This function is shown in Figure 9.7.

We first note that this function is not zero for $t < 0$. Consequently, we must shift it to the right by an amount 2 as per the instructions in step 1. That is to say, we form a new function $g(t) = f(t - 2)$. This is shown in Figure 9.8.

Step 2 requires that we sample the function with a sampling rate $\Delta t = 1/10\Omega$ and choose the number of samples such that

$$N\Delta t > 10(T + A) = 10(2 + 2) = 40. \tag{9.6}$$

As is most often the case, we don't know the value of the band limit Ω a priori, and therefore we have to guess at a value of Δt by an examination of the time domain function $g(t)$. Basically, we choose a sampling rate such that a smooth curve passed through the sampled sequence values will "strongly resemble" the function. In this case a value of $\Delta t = 0.08$ is sufficient, as illustrated in Figure 9.9. Using this rate and equation (9.6), we determine the number or samples to be equal to 500.

Next (step 3) we calculate $\{G(j)\}$, the discrete Fourier transform of the sequence $\{g(k)\}$, and multiply the terms by $N\Delta t$ to obtain the values of the sequence $\{G(j\Delta w)\}$ shown in Figure 9.10. Note that this transform is complex valued with a real and imaginary portion.

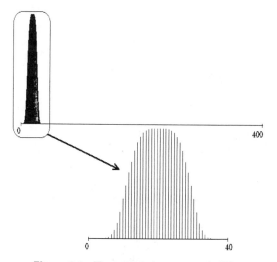

Figure 9.9   The sampled sequence $\{g(k)\}$.

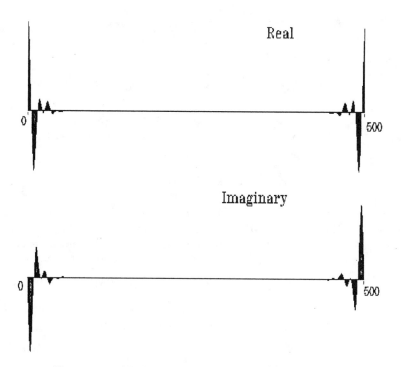

Figure 9.10 $\{G(j)\}$, the discrete Fourier transform of $\{g(k)\}$.

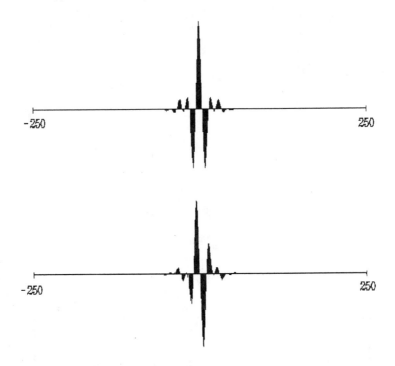

Figure 9.11 Shifted discrete Fourier transform sequence.

As per the instructions in step 4, we use Theorem 8.3 to obtain values for the negative indices:

$$G(-j) = G(N - j).$$

The results of this move are shown in Figure 9.11.

In step 5 we recover $G(w)$, the transform of $g(t) = f(t - 2)$, by passing a smooth curve between the values of the sequence $\{G(j)\}$. This is shown in Figure 9.12. Note that the frequency scale is generated as per the formula $\Delta w = 1/N\Delta t = 1/40$. Finally, the desired result $F(w)$ is obtained, as per step 6, by multiplying $G(w)$ by

$$e^{2\pi i w A} = e^{2\pi i w 2} = \cos(4\pi w) + i \sin(4\pi w).$$

This is shown in Figure 9.13. Note that this transform is real and even, as it certainly must be since $f(t)$ is a real and even function (see Theorem 6.16).

COMPUTER-GENERATED FOURIER SERIES

In this section we extend the results of the previous section and present a method that permits the digital calculation of the Fourier series coefficients of a (periodic) function. Just as in the previous section, we accomplish this by

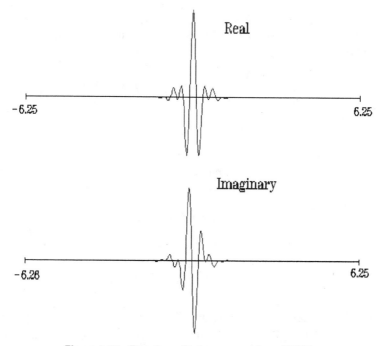

Figure 9.12 Transform $G(w)$ recovered from $\{G(j)\}$.

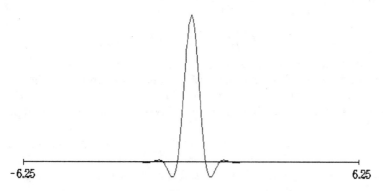

Figure 9.13 Digitally obtained Fourier transform $F(w)$.

sampling the function and obtaining the desired coefficients from the terms of the discrete Fourier transform of the sampled sequence $\{f(k)\}$.

We begin with a function $f(t)$ that is assumed to be periodic over the interval $T = [-T/2, T/2]$. In this case we sample the function with sampling rate $\Delta t = 1/2\Omega$ over the interval T and choose the number of samples N such that $N\Delta t = T$.

In Chapter 7 (equation (7.37)) we saw that the (complex) Fourier series coefficients of the function $f(t)$ could be obtained from the Fourier transform of the function at equally spaced intervals $w = j\Delta w = j/T$—in other words, at values of w equal to $j\Delta w = j/N\Delta t$. We now use this fact along with equation (9.4) (with $w = j/N\Delta t$) to obtain

$$C_k = F(j\Delta w) = (1/T)(1/2\Omega) \sum_{k=0}^{N-1} f(k)e^{-2\pi i k j \Delta w \Delta t}.$$

However,

$$1/T = 1/N\Delta t,$$

$$1/2\Omega = \Delta t,$$

and

$$\Delta w \Delta t = 1/N.$$

Thus, the previous equation becomes

$$C_k = \frac{1}{N} \sum_{k=0}^{N-1} f(k)e^{-2\pi i k j/N}. \tag{9.7}$$

Therefore, we see that the (complex) Fourier series coefficients are equal to the terms of the discrete Fourier transform of the sequence $\{f(k)\}$ when the sampling rate is $\Delta t = 1/2\Omega$, and the number of samples N are chosen such that $N\Delta t = T$. Note that the values of the coefficients for negative indices $(-j)$ are obtained by means of Theorem 8.3, that is,

$$C_{-j} = C_{N-j} = F(N - j).$$

It is important to note that although there are an infinite number of Fourier series coefficients C_j, the method just presented will only yield the first N of them. However, inasmuch as all real-world functions are band limited, we can always choose a sufficiently large value of N to give a reasonable representation of the function.

We also note that if Ω is a band limit of the function, then so too is 5Ω. Thus, $\Delta t = 1/10\Omega$ will also work in this scheme as long as N is chosen such that $N\Delta t = 1/10\Omega$. In actual practice, this is the preferred sampling rate.

SUPER-GAUSSIAN WINDOWS

When we sample a function we must concern ourselves with what is known as *window effects* or *windowing*. This is essentially a truncation of the function. That is to say, the Fourier transform, as given per equation (6.1), is defined over the entire real line $R = (-\infty, \infty)$. However, practical limitations (such as memory size of the digitizing and recording equipment) often make it impossible to properly sample a function over its entire domain. For example, in Figure 9.14 we show a function that has bounded support over the domain

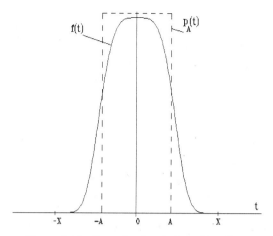

Figure 9.14 Example of a windowed function.

$[-X, X]$ but that, for practical reasons, has only been sampled over the domain $[-A, A]$. In this case the result is a sharp cutoff or discontinuity at the endpoints $-A$ and A. Mathematically, we describe windowing as the product of the function $f(t)$ and a pulse function $p_A(t)$, that is,

$$g(t) = f(t)p_A(t).$$

In other words, instead of the function $f(t)$, because of window effects, we have the function $g(t)$. In the frequency domain we obtain

$$G(w) = 2AF(w) * \text{sinc}(2\pi Aw), \tag{9.8}$$

where $F(w)$ is the desired transform and $G(w)$ is the smeared transform that results from the convolution of the original function transform $F(w)$ and the transform of the window which is given as $2A\,\text{sinc}(2\pi wA)$ (see the product theorem (Theorem 6.18)). Using the integral expression for convolution as given by equation (6.28), we are able to write

$$G(w) = 2A \int_{-\infty}^{\infty} F(x)\,\text{sinc}[2\pi A(w - x)]\,dx.$$

In Figure 9.15 we show both $F(x)$ and $\text{sinc}[2\pi A(w - x)]$. As we learned in Chapter 6, the value of the convolution function $G(w)$ (for a particular value of w) is equal to the area under the product curve of the two functions being convolved. As can be appreciated from this figure, the resulting product is dependent upon both the separation w and the relative widths of $F(w)$ and $\text{sinc}[2\pi Aw]$. Because of these factors, we now begin to appreciate why there is no general theory of windowing. We can only look for trends. For example, if the window is wide (i.e., A is large) relative to X, then the window problems will be less. As a matter of fact, in the limit as A approaches infinity,

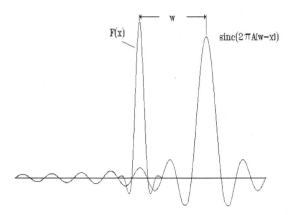

Figure 9.15 Effects of windowing considered in the frequency domain.

sinc$[2\pi Aw]$ becomes a delta function of strength $1/2A$. In this case, equation (9.8) becomes

$$G(w) = 2AF(w)*\delta(w)/2A = F(w)*\delta(w) = F(w),$$

and, as expected, there are no windowing problems at all.

To help illustrate these concepts let us choose a specific function, namely, the Gaussian

$$f(t) = \exp[-t^2], \qquad t \in [-3.0, 3.0],$$

and place three different windows over it. The first window (see Figure 9.16a) is such that it clips, or apertures, the function at the 0.25 level. That is to say, the half-width A of the window is chosen such that

$$F(A) = \exp[-A^2] = 0.25, \quad \text{or} \quad A = 1.177.$$

Shown in Figure 9.16b is convolution of the true transform $F(w)$ and the sinc$(2\pi wA)$ function for a particular value of w. Finally, in Figure 9.16c we show (on a different scale) the transform $G(w)$, that is, the convolution of $F(w)$ and sinc$(2\pi wA)$. In Figures 9.17 and 9.18 we repeat this exercise for two additional windows that clip the function at the 0.10 ($A = 1.517$) and 0.005 ($A = 2.302$) levels. When comparing these figures, the important thing to note is the relative widths of the two transforms $F(w)$ and sinc$(2\pi wA)$ and how this correlates to the resulting transform $G(w)$.

As may or may not be evident from the previous discussions, the sharp discontinuities at the window edges ($|t| = A$) give rise to the ringing in the transform $G(w)$. Even though it may not be practical to capture the entire function due to physical limitations, we can reduce the severity of the ringing by reducing the sharpness of the cutoff value. This is accomplished by placing an artificial window over the function that *gradually* reduces the function to zero at the endpoints. If figurative terms, we use an artificial window with

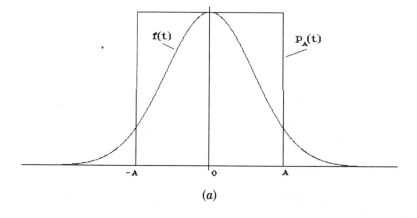

(a)

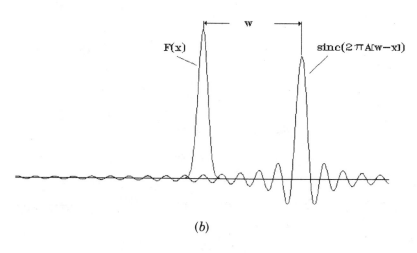

(b)

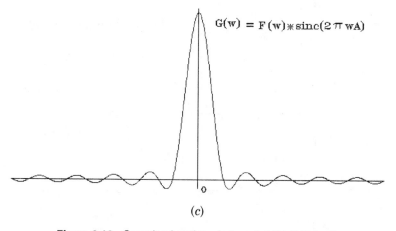

(c)

Figure 9.16 Gaussian function windowed at the 0.25 level.

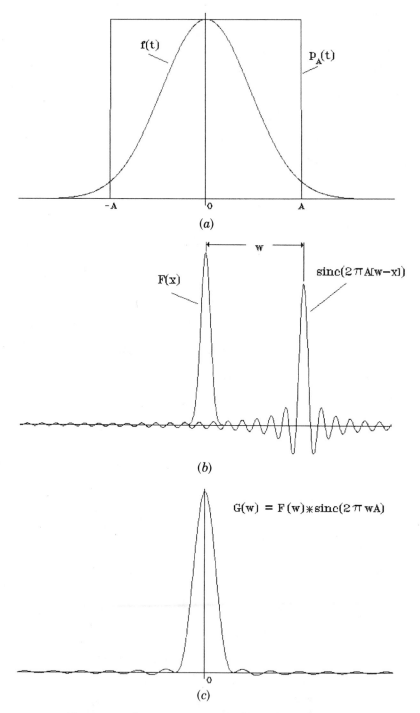

Figure 9.17 Gaussian function windowed at the 0.1 level.

rounded corners. Several of the more well-known windows are the Kaiser, Lanczos, exponential, hamming, and hanning. Another very effective and versatile window is the super-Gaussian. Inasmuch as this window is not covered in the literature in the depth that the other mentioned ones are, we end this chapter with a discussion of the super-Gaussian functions.

Mathematically, we define the super-Gaussian function, or simply the SG function, as

$$SG(t; A, N) = \exp[-|t/A|^N]. \tag{9.9}$$

As can be seen from equation (9.9), there are two parameters in the expression for the SG function, namely, A and N. N is called the power of the function, and A is known as the $(1/e)$ radius of the function. For large values of N, A is also referred to as the half-width of the function.

In Figure 9.19 we show the SG function for several different values of the power N. In this figure we note that they all cross at $t = A$, that is,

$$SG(A; A, N) = \exp[-|A/A|^N] = e^{-1} = 1/e = .367879.$$

Thus, we see the logic behind calling A the $1/e$ radius of the SG function. When $N = 2$ the SG function is a simple Gaussian function, that is,

$$SG(t; A, 2) = \exp[-(t/A)^2]. \tag{9.10}$$

On the other hand, in the limit as N approaches infinity, the SG function becomes a pulse function of half-width A, that is,

$$\lim_{N \to \infty} SG(t; A, N) = p_A(t). \tag{9.11}$$

For values of N between 2 and ∞ the SG functions may be considered pulse functions with rounded corners.

We note that for all *finite* powers of N, the SG functions roll off toward zero but never exactly equal zero. In other words, at the edge of the window (denoted as $t = T$) the SG function will not be exactly equal to zero but, instead, will be equal to a (small) finite value ε. Mathematically, we have

$$SG(T; A, N) = \varepsilon,$$

or

$$\exp[-|T/A|^N] = \varepsilon,$$

or

$$T/A = [\ln(1/\varepsilon)]^{-1/N}.$$

For most practical applications, a reduction factor of $1/\varepsilon = 1000$ is usually sufficient. Thus, the A/T ($1/e$ radius to edge) ratio becomes

$$A/T = (0.1448)^{1/N}. \tag{9.12}$$

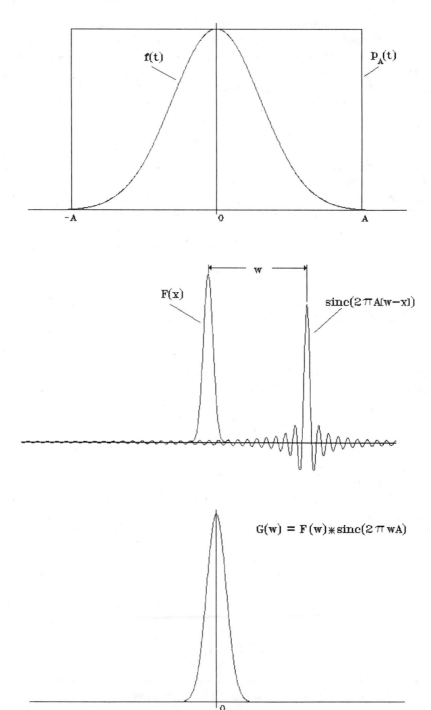

Figure 9.18 Gaussian function windowed at the 0.005 level.

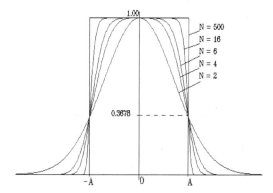

Figure 9.19 Super-Gaussian function for various values of the power N.

For example, if the window edge for a 5-power super-Gaussian function is specified as $T = 2$, then to obtain a reduction factor of 1000 at this edge the $1/e$ radius A of the filter must be chosen as

$$A = 2(.1448)^{1/5} = 1.359.$$

When windowing a function, we obviously wish to smoothly reduce the function to zero at the window edge. However, we wish to accomplish this with the least amount of disruption to the overall function. For example, consider the function $f(t)$ shown in Figure 9.20 over the domain $[0,6]$. Let us assume that we are to sample this function with sampling rate $\Delta t = 0.015$ (in order to determine its Fourier transform using digital techniques). Also, let us suppose that our sampling and recording system can only accommodate 128 samples because of memory limitations. Clearly, $N\Delta t = 1.92 < 6$. Thus, we must "chop this function off" at $T = 1.92$. A rectangular, or square, window of half-width $A = 1.92$ will cause a sharp discontinuity at the window edge as shown in Figure 9.21. On the other hand, the rest of the function over the domain $[0,T)$ is left unchanged.

Now suppose we use a Gaussian function to smoothly reduce the function (by a factor of 1000) at the window edge. Using equation (9.12) with $N = 2$ and $T = 1.92$, we determine the $1/e$ radius A to be equal to 0.7306. This window is shown along with the function in Figure 9.22a, and the resulting product (or windowed function) is shown in Figure 9.22b. Certainly the function has been gradually reduced to zero, but it has also been severely altered over the rest of the domain $[0,T)$.

Clearly, values of N between these two extremes offer a compromise between edge roll-off and overall function preservation. Shown in Figures 9.23 and 9.24 is the function windowed by using a 4- and 6-power super-Gaussian function, respectively.

As we learned earlier in this section, to appreciate the effect of a window on the Fourier transform of a function $f(t)$ we must examine the convolution

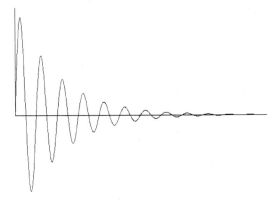

Figure 9.20   Example of function to be windowed.

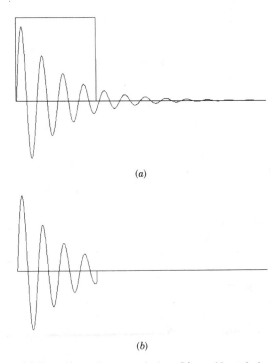

(a)

(b)

Figure 9.21   (a) Function and square window; (b) resulting windowed product.

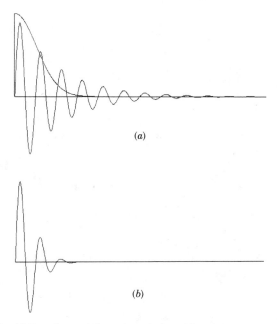

Figure 9.22   (a) Function and Gaussian window; (b) resulting windowed product.

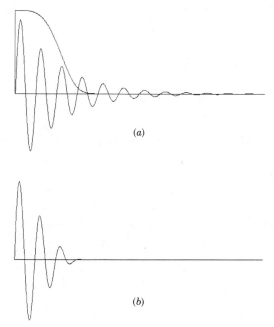

Figure 9.23   (a) Function and 4-power super-Gaussian window; (b) resulting windowed product.

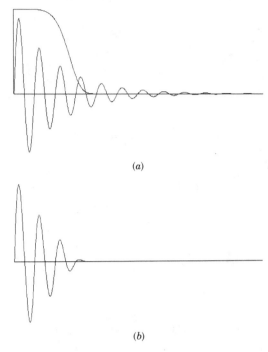

(a)

(b)

Figure 9.24 (a) Function and 6-power super-Gaussian window; (b) resulting windowed product.

product of $F(w)$ with the Fourier transform of the window function. Therefore, we now consider the Fourier transform of the super-Gaussian functions.

For the two limiting cases of $N = 2$ and $N = \infty$ we are able to write analytical expressions for the transforms, that is,

$$\mathcal{F}[SG(t; A, 2)] = \pi^{1/2} A \exp[(-w/B)^2], \qquad \text{where} \quad B = 1/\pi A, \quad (9.13)$$

$$\mathcal{F}[SG(t; A, \infty)] = 2A \operatorname{sinc}(2\pi w A). \qquad (9.14)$$

These transforms are shown in Figure 9.25. We note that the low-frequency portion of these transforms are very similar. The $N = 2$ transform has a peak ($w = 0$) value of $\pi^{1/2} A = 1.772A$, and the $N = \infty$ transform has a peak value of $2A$. In other words, they only differ by approximately 12%. Also, the $1/e$ radius of the $N = 2$ transform is $1/\pi A$, while that of the $N = \infty$ transform is $1.09/\pi A$. Thus, they only differ by approximately 9%. The significant difference between the two transforms is the ringing exhibited by the $N = \infty$ super-Gaussian transform.

The transform for values of N between 2 and ∞ must be digitally obtained. In Figure 9.26 we show three such transforms (for $N = 4$, 6, and 16). Note that as the power N increases, so too does the ringing and/or amount of higher frequency present in the transform.

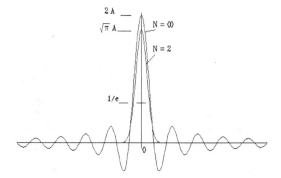

Figure 9.25 Fourier transform of the $N = 2$ and $N = \infty$ super-Gaussians.

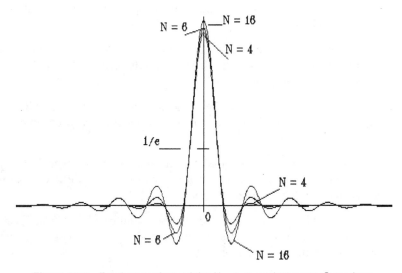

Figure 9.26 Fourier transform of the $N = 4$, 6, and 16 super-Gaussians.

SUMMARY

In this chapter we used the comb function to develop a description of sampling. We showed that when a function was sampled using the Nyquist rate, the function could be uniquely and exactly recovered from its sampled sequence. We then showed how the Fourier transform (as well as the Fourier series coefficients) of a function could be obtain from the discrete Fourier transform of the sequence generated by sampling the function. Finally, we discussed the super-Gaussian functions, which make ideal windows to be used when sampling a function.

PROBLEMS

1 In the following problems you will be asked to perform the digital calculation of both the Fourier transform and the Fourier series coefficients of various functions. Either write your own computer program that performs the six-step transform operation presented in this chapter or key into your computer the program provided in the appendix. (*Note:* This program requires that the number of data points, N, be a power of 2.)

2 Use the computer program generated in problem 1 and choose a proper sampling rate and number of samples to digitally obtain the Fourier transform of a Gaussian function with $1/e$ radius equal to 1 (see equation (6.9)). Check your answer by plotting it along with the analytical expression given by equation (6.10). Cut your sampling rate in half and see if your results change.

3 Same as problem 2, but use the one-sided exponential function given by equation (6.7) with $a = 2$. Check your results against the analytical expression given by equation (6.8).

4 Same as problem 2, but use the pulse function with a half-width $a = 2$ (see equation (6.5)). Check your results using the analytical expression given by equation (6.6).

5 Same as problem 2, but use the shifted pulse function given in Chapter 6, Example 5. In this exercise let $b = 2$ and $a = .5$. Check your answer against the results presented in equation (6.11).

6 Same as problem 2, but use the damped sinusoidal function given by equation (6.12). In this exercise let $\sigma = 2$ and $w_0 = 4$. Check your answer against the results given equation (6.13).

7 Same as problem 2, but use the unit step function given in equation (7.12). Compare your results with those given by equation (7.14). (*Note:* In this problem you will have to apply a window to the function in order to reduce the ringing.) Experiment with different windows to optimize your answer.

8 Use the techniques presented in this chapter (and the computer program of problem 1) to determine the first 50 Fourier series coefficients (complex form) of the triangle function given by equation (5.56) over the domain $[-1, 1]$. Convert these results to the rectangular form (using Euler's equations presented in Chapter 5) and check your results against those given in Chapter 5, Example 1.

9 Same as problem 8, but use the function $f(t) = t^2$ over the domain $[-2, 2]$. Use the results of Example 4, Chapter 5 to check your results.

10 Same as problem 8, but use the function $f(t) = t$ over the domain $[-3,3]$. Use the results of Example 5, Chapter 5 to check your results.

11 Same as problem 8, but use the boxcar function shown in Figure 1.3. Check your answers against those given in Chapter 1.

12 Use the computer program generated in problem 1 to calculate the Fourier transform of the super-Gaussian functions for values of N ranging from 3 to 30. Use a $1/e$ radius of $A = 1$.

13 Use the computer program of problem 1 and calculate the Fourier transform of a triangle function of half-width $2a$. Next calculate the Fourier transform of a pulse function of half-width a and square the results. Do they agree? (See the convolution theorem (Theorem 6.17)). In this problem use $a = 2$.

14 Using equation (9.12) and a window edge of $T = 1.92$, what are the $1/e$ radii of the $N = 4$ and $N = 6$ super-Gaussian windows used in Figures 9.23 and 9.24?

BIBLIOGRAPHY

Hamming, R. W., *Digital Filters*, Prentice-Hall, Englewood Cliffs, N.J., 1977.

Papoulis, A., *Systems and Transforms with Applicatons in Optics*, McGraw-Hill, New York, 1968.

Rabiner, L., and B. Gold, *Theory and Applications of Digital Signal Processing*, Prentice-Hall, New York, 1975.

Weaver, H. J., *Applications of Discrete and Continuous Fourier Analysis*, John Wiley & Sons, New York, 1983.

APPENDIX

FOURIER TRANSFORM FORTRAN SUBROUTINE

The following program (written in FORTRAN) will calculate the Fourier transform of a function that has been properly sampled and converted to a sequence by the rules specified in Chapter 9. All of the zero shifting and phase adjustments described in that chapter are performed in this program.

```
        subroutine fourier(x,fr,fi,n)
        dimension x(1),fr(1),fi(1)
        dimension dummy(2048)
c...
c...
c...
c...                    F O U R I E R    T R A N S F O R M
c...                         S U B R O U T I N E
c...
c...                              Coded by
c...
c...                        H. Joseph Weaver, Ph.D.
c...
c...
c...
c...    This subroutine will take in the variable arrays x, fr, and fi,
c...    where x is the independent variable, fr is the real portion of
c...    the dependent variable, and fi is the imaginary portion of the
c...    dependent variable. That is to say, we are dealing with the
c...    complex function fr(x)+ifi(x). The program will return the
c...    Fourier transform of this function in these same arrays; x will
c...    now contain the frequency, and fr is the real portion of the transform.
c...
c...    n is the dimension of the previously discussed arrays, and
c...    it must be equal to a power of 2 (i.e. 2, 4, 8, 16, 32, 64,...).
```

```
c...
c...   Note: If the function to be transformed is real, then the
c...   elements in the imaginary array should be set to zero.
c...
c...
c...   The array DUMMY is just that. It is used in the
c...   subroutine SHFT. Its dimension limits the size of the
c...   transform that can be performed.
c...
c...   This subroutine uses the fast Fourier transform algorithm
c...   coded by Bruce Langdon. It is based upon a publication by
c...   R. C. Singleton.
c...
c...
c...
       pi=4.*atan(1.)
       incp=1
       isign=-1
       call ffts(fr,fi,n,incp,isign)
c...
c...
c...   Shift function so frequencies read from -wmax to +wmax
       call shft(x,fr,fi,n,dummy)
c...
c...
c...   Determine the amount by which functions are shifted (if any)
c...   shft1 is distance from 0 location to x(1)
       call zeros(x,n,shft1)
c...
c...   Generate frequency scale for function
       dx=x(2)-x(1)
       dw1=1./(dx*n)
       n2=n/2
       do 10 i=1,n2
       x(n2+i)=(i-1)*dw1
    10 x(n2+1-i)=-i*dw1
c...
c...   Scale function
       do 20 i=1,n
       fr(i)=fr(i)*dx
    20 fi(i)=fi(i)*dx
c...
c...
c...   Compensate for any phase shift due to zero location
       if(shft1.le.0.) go to 99
       do 30 i=1,n
       xx=fr(i)
       yy=fi(i)
       arg=2.*pi*x(i)*shft1
       sf=sin(arg)
       cf=cos(arg)
       fr(i)=xx*cf-yy*sf
    30 fi(i)=yy*cf+xx*sf
c...
```

```
      99 continue
         return
         end
         subroutine ffts(r,i,n,incp,signp)
c     - - - - - - - - - - - - - - - - - - - - - - - - - - - - - - - - - - - - - - - - - - - - - -
c...   programmed by a.b. langdon
c...   periodic complex fourier transform
c     - - - - - - - - - - - - - - - - - - - - - - - - - - - - - - - - - - - - - - - - - - - - - -
         real r(1),i(1)
         integer signp,span,rc
         real sines(15),io,i1
         data sines(1)/0./
         if(sines(1) .eq. 1) go to 1
         sines(1)=1.
         t=atan(1.)
         do 2 is=2,15
         sines(is)=sin(t)
       2 t=t/2.
       1 continue
         if(n .eq. 1) return
         inc=incp
         sgn=isign(1,signp)
         ninc=n*inc
         span=ninc
         it=n/2

         do 3 is=1,15
         if(it .eq. 1) go to 12
       3 it=it/2
      10 t=s+(so*c-co*s)
         c=c-(co*c+so*s)
         s=t
      11 k1=ko+span
         ro=r(ko+1)
         r1=r(k1+1)
         io=i(ko+1)
         i1=i(k1+1)
         r(ko+1)=ro+r1
         i(ko+1)=io+i1
         ro=ro-r1
         io=io-i1
         r(k1+1)=c*ro-s*io
         i(k1+1)=s*ro+c*io
         ko=k1+span
         if(ko .lt. ninc) go to 11
         k1=ko-ninc
         c=-c
         ko=span-k1
         if(k1 .lt. ko) go to 11
         ko=ko+inc
         k1=span-ko
         if(ko .lt. k1) go to 10
      12 continue
         span=span/2
```

```
      ko=0

13 k1=ko+span
      ro=r(ko+1)
      r1=r(k1+1)
      io=i(ko+1)
      i1=i(k1+1)
      r(ko+1)=ro+r1
      i(ko+1)=io+i1
      r(k1+1)=ro-r1
      i(k1+1)=io-i1
      ko=k1+span
      if(ko .lt. ninc) go to 13

      if(span .eq. inc) go to 20
      ko=span/2
14 k1=ko+span
      ro=r(ko+1)
      r1=r(k1+1)
      io=i(ko+1)
      i1=i(k1+1)
      r(ko+1)=ro+r1

      i(ko+1)=io+i1
      r(k1+1)=sgn*(i1-io)
      i(k1+1)=sgn*(ro-r1)
      ko=k1+span
      if(ko .lt. ninc) go to 14
      k1=inc+inc
      if(span .eq. k1) go to 12
      co=2.*sines(is)**2
      is=is-1
      s=sign(sines(is),sgn)
      so=s
      c=1.-co
      ko=inc
      go to 11
20 n1=ninc-inc
      n2=ninc/2
      rc=0
      ji=0
      ij=ji
      if(n2 .eq. inc) return

      go to 22
21 ij=n1-ij
      ji=n1-ji
      t=r(ij+1)
      r(ij+1)=r(ji+1)
      r(ji+1)=t
      t=i(ij+1)
      i(ij+1)=i(ji+1)
      i(ji+1)=t
      if(ij .gt. n2) go to 21
```

```
   22 ij=ij+inc
      ji=ji+n2
      t=r(ij+1)
      r(ij+1)=r(ji+1)
      r(ji+1)=t
      t=i(ij+1)
      i(ij+1)=i(ji+1)
      i(ji+1)=t
      it=n2
   23 it=it/2
      rc=rc-it
      if(rc .ge. 0) go to 23
      rc=rc+2*it
      ji=rc
      ij=ij+inc
      if(ij .lt. ji) go to 21
      if(ij .lt. n2) go to 22
      return
      end
      subroutine shft(x,yr,yi,n,d3)
      dimension x(1),yr(1),yi(1)
      dimension d3(1)
c...
c...
c...  this subroutine will perform the post transform shifting
c...  to obtain negative frequencies
c...
c...  real first
      n2=n/2
      do 2 i=1,n2
      d3(n2+i)=yr(i)
    2 d3(i)=yr(n2+i)
      do 3 i=1,n
    3 yr(i)=d3(i)
c...
c...  now imaginary
      do 4 i =1,n2
      d3(n2+i)=yi(i)
    4 d3(i)=yi(n2+i)
      do 5 i=1,n
    5 yi(i)=d3(i)
c...
c...
      return
      end
      subroutine zeros(x,n,shift)
      dimension x(1)
c...
c...  this subroutine will determine how far the zero location of
c...  the function is shifted from x(1).
c...
c...
c...  if x(1) is greater than 0 then assume no shift.
      if (x(1).gt.0.) shift=-2.
```

```
          if(x(1).gt.0.) return
c...
c...
          iz=0
          do 1 i=1,n
          iz=i
          if(x(i).ge.0.) go to 2
        1 continue
c...
c...   if code makes it to here then assume there is no shift
          shift=-2
          return
c...
c...
        2 continue
          if(x(iz).eq.0.) shift=-x(1)
          if(x(iz).eq.0.) return
c...
          shift=.5*(x(iz)+x(iz-1))-x(1)
c...
          return
          end
```

INDEX

Abelian group, 29
Absolutely integrable function, 154
Accumulation point, 40
Aliasing, 273
Almost everywhere, 54
Associativity, 26
Autocorrelation, 196
  Fourier transform, 196

Band limited function, 270
Beppo–Levi Theorem:
  for sequences, 66
  for series, 67
Bessel's inequality, 106, 114
Bounded sequence, 43, 237
Bounded support, 162, 230
Boxcar function, 5, 19, 232
  Fourier series, 8
  Fourier transform, 232

Cartesian product, 24
Cauchy Convergence, 47
  of step functions, 58
  uniform, 48
Closed neighborhood, 39
Comb distribution, 89, 221
  Fourier transform, 223
Commutativity, 26
Complement of 2 sets, 23
Complete self-reciprocity, 159

Complete set, 48, 107
Complete system, properties, 108
Completeness condition, 107
Complex conjugate:
  of distribution, 220
  of function, 174, 194
    Fourier transform, 196
  of sequence, 254
Complex exponential form of function, 11
Complex exponential Fourier series
  representation, 102, 103
Complex exponential system of orthogonal
  functions, 115
Complex function:
  Fourier series, 130
Complex number system:
  as vector space, 31
  inverse element, 49
  norm, 36
Computer generated Fourier series, 280
Computer generated Fourier transform, 274
  six step procedure, 277
Continuity:
  of Fourier transform, 162
  of functions, 41
Convergence in the mean, 107, 114
Convergence:
  of sequence, 43
    of Lebesgue functions, 65
  of sequence of step functions, 58

Convolution:
  of equal pulse function, 184
  existence, 225
    properties, 258
  of two distributions, 224
  of two functions, 181
  of two Gaussian functions, 193
  of two sequences, 257
  of unequal pulse functions, 191
  properties, 181–183
  using shifted delta distribution, 226
Convolution theorem, 189
  for discrete Fourier transform, 258
  for distributions, 226
Cooley–Tukey algorithm, 263
Cosine function:
  Fourier transform, 214
  sum of two angles, 111
Countable set, 35
Cross-correlation, 194
  Fourier transform, 196
  properties, 195

Damped sinusoid, 165
  Fourier transform, 167
Deleted neighborhood, 39
Delta distribution (function), 17, 86
  derivative, 95
  Fourier transform, 19
  unit element under convolution, 226
Delta sequence, 250
  discrete Fourier transform of, 250
Derivative:
  of distribution, 87, 91, 214
  Fourier transform, 171
  at jump discontinuity, 92, 96
Derivative of Fourier transform, 170, 215
Difference of two distributions, 88
Dirac function, 17
  Fourier transform, 19
Dirichlet kernel function, 117
Discrete Fourier transform, 237
  convolution theorem, 258
  of delta sequence, 250
  of even sequence, 254
  extending the domain, 242
  first forward difference, 252
  first shifting theorem, 245
  of imaginary sequence, 256
  linearity, 244
  of odd sequence, 254
  periodicity, 242
  product theorem, 259
  real sequence, 255

reciprocity, 242
second shifting theorem, 247
symmetry considerations, 253
transform of a transform, 249
of unit constant sequence, 250
Distance function, 35
Distribution, 85
  convolution product, 224
  derivative, 91
  odd and even, 94
  properties, 87
Distributivity, 26
Domain of function, 33

Empty set, 22
Equality of two distributions, 88
Equivalent sets, 34
Euler's equations, 115
Even distribution, 94, 219
Even function, 176
  Fourier transform, 176
Even sequence, 254
  discrete Fourier transform of, 254
Existence of convolution product, 225
Existence of Fourier transform, 154
Exponential function (one-sided), 11
  Fourier transform, 11
Extending domain of discrete Fourier
    transform, 242

Fast Fourier transform, 263
Field, 29
Finite sequence, 34
Finite set, 22
First forward difference, 251
  discrete Fourier transform, 252
  of discrete Fourier transform, 252
First shifting theorem:
  for discrete Fourier transform, 245
  for Fourier series, 132
  for Fourier transforms, 164, 211
Fourier series, 3
  approximation, 5
  of boxcar function, 8
  coefficients, 102, 104, 113
  of complex functions, 130
  of even function, 136
  first shifting theorem, 132
  linearity, 131
  of odd function, 138
  of one-sided exponential function, 148
  of parabolic function, 144
  of ramp function, 144
  of rectified cosine function, 147

of triangle function, 5, 139
  with hole at origin, 139
  second shifting theorem, 133
  of "V" function, 141
Fourier transform, 11, 153
  of autocorrelation function, 196
  behavior at infinity, 160, 172
  of boxcar function, 232
  of comb distribution, 223
  of complex conjugate, 196, 220
  continuity, 162
  convolution theorem, 187
  of cosine function, 214
  of cross-correlation function, 196
  of cross-correlation function symmetry, 197
  of damped sinusoid, 167
  of delta distribution, 209
  of derivative, 171
  of derivative of exponential function, 217
  of derivative of pulse function, 217
  of distribution, 207
  of even function, 176
  existence, 154
  of exponential function, 11
  first shifting theorem, 164, 211
  of $f(t)\cos(2\pi at)$, 212
  of $f(t)\sin(2\pi at)$, 212
  of Fourier transform, 173
  of Gaussian function, 156, 163
  of Hermite functions, 198
  of imaginary function, 175
  linearity of, 159, 210
  of odd function, 176
  of one-sided exponential function, 156, 163
  Parseval's energy formula, 197
  of periodic function, 230
  product theorem, 193
  of pulse function, 12, 155, 163
  of real function, 174, 175
  reciprocity of, 178
  scale change, 167, 211
  second shifting theorem, 165, 212
  of sgn function, 218
  of shifted delta distribution, 213
  of sinc function, 214
  of sine function, 214
  of super-Gaussian function, 292
  symmetry consideration, 174, 177, 220
  of tophat function, 12
  of triangle function, 192
  uniqueness of, 178
  of unit constant function, 213
  of unit step function, 218
Frequency of Gibb's ringing, 127

Function, 32, 34
Function of bounded support, 162, 230
Function with hole at origin, 139
  Fourier series of, 139

Gaussian function, 156, 163, 172
  Fourier transform of, 158
General system of orthogonal functions, 103
Gibb's effect, 8, 125
  frequency of ringing, 127, 145
Group, 28

Hermite functions, 199
  Fourier transform of, 203
Hermite polynomials, 198
  recursion formulas, 201

Imaginary distribution, 94, 220
Imaginary sequence, 254
  discrete Fourier transform, 256
Impulse function, 17, 86
  Fourier transform, 19
  as limit, 17
Impulse response, 228
Infinite sequence, 35
Infinite set, 22
Inner product, 81
  existence considerations, 83
  properties, 82
Integral functions, 72
Intersection of sets, 22
Invariant system, 228
Inverse discrete Fourier transform, 237
Inverse element of set, 27
Inverse Fourier transform, 11

Laws of internal composition, 25
Lebesgue Convergence Theorem, 68, 69
Lebesgue covering, 52
Lebesgue function, 61
  on unbounded interval, 69
Lebesgue measure, 51, 53
Limit:
  of function, 41, 42
  of sequence, 42
Linear system, 228
Linearity:
  of discrete Fourier transform, 244
  of Fourier series, 131
  of Fourier transform, 159, 210
L^{-1} Norm of function, 162

Mean square error, 105, 114
Metric space, 35

Monotonic decreasing sequence, 43
Monotonic increasing sequence, 43
Multiplication of sequence by constant, 237

Neighborhood, 39
Norm, 36
  of orthogonal function, 103
  of step function vector space, 57
Normal distribution, 157
  Fourier transform, 158
Nth-order sequence, 235
  product, 236
  sum, 236
nth partial sum of series, 105
Null distribution, 86
Nyquist sampling rate, 273

Odd distribution, 94, 219
Odd function, 176
  Fourier transform of, 176
Odd sequence, 254
  discrete Fourier transform of, 254
Odd/even function, 136
  Fourier series, 136
  Fourier transform, 176
One-sided exponential function, 156, 163,
  170, 171, 174, 175
  derivative, 216
  Fourier series, 148
  Fourier transform, 156
One-to-one function, 34
Open interval, 52
Open neighborhood, 39
Ordered pairs, 24
Ordered relation, 37
Orthonormal system of functions, 103
  complex exponential, 115
  trigonometric, 111

Parabolic function, 143
  Fourier series of, 144
Parseval's equation, 107, 144, 197
Periodic function, 101
  as convolution, 230
  Fourier transform, 230
Periodicity of discrete Fourier transform, 242
Physical interpretation of convolution, 227
Pointwise convergence:
  of Fourier series, 121
  of sequences, 43
Principle value of function, 154
Product of distribution and complex number,
  87

Product of distribution and function, 88
Product of sine and cosine, 111
Product theorem:
  for discrete Fourier transform, 259
  for Fourier transform, 193, 226
Product of Nth order sequences, 236
Proper subset, 22
Pulse function, 12, 155, 163, 177, 227
  derivative, 217
  Fourier transform, 12, 155
Pyramid function, 187

Ramp function, 20, 144
  Fourier series, 144
Range of function, 33
Real distribution, 94, 219
Real function, 174
  Fourier transform, 174
Real sequence, 254
  discrete Fourier transform, 255
Real world sampling rate, 274
Reciprocity:
  of discrete Fourier transform, 242
  of Fourier transform, 178
Rectified cosine function, 145
  Fourier series, 147
Recursion formulas for Hermite polynomials,
  198
Relation, 32
Riemann Integral, 64
Riemann Lebesgue lemma, 116
Riemann localization theorem, 123
Ring, 49

Sampling rate, 270
  Nyquist, 272
  real world, 274
Scale change:
  of distribution, 89
  for Fourier transforms, 167, 211
Second shifting theorem:
  for discrete Fourier transform, 247
  for Fourier series, 133
  for Fourier transform, 165, 212
Self-reciprocity, 159
Sequence, 34
  Nth order, 235
Set, 21
Set of measure zero, 54
Sgn function, 218
  Fourier transform of, 218
Shifted delta distribution:
  convolution, 226
  Fourier transform, 213

Simple convergence of sequence, 43
Simultaneous calculation of two discrete
   transforms, 260
Sinc function, Fourier transform, 214
Sine function, sum of two angles, 111
Square-summable functions, 76
Stationary system, 228
Step function, 54
Subset, 22
Sum of Nth order sequences, 236
Sum of two distributions, 87
  symmetry considerations, 174
Super-Gaussian functions, 287
  Fourier transform, 291
Symmetry considerations:
  discrete Fourier transform, 253, 256
  Fourier transform, 174, 219
    of cross-correlation function, 196
System, 228

Tophat function, 12
  Fourier transform, 12
Totally ordered set, 38
Transform of transform:
  discrete Fourier transform, 249
  Fourier transform, 173, 213
Transform of derivative (distribution), 216
Translation of distribution, 89
Triangle function, 3, 19, 20, 138, 188

as convolution of two pulse functions, 185
Fourier series, 5, 139
Fourier transform, 188
with hole at origin, 139
Trigonometric orthogonal system, 111, 113

Uniform continuity, 42
Uniform convergence of sequence, 44
Union of sets, 22
Uniqueness of the Fourier transform, 178
Unit constant function, 227
  Fourier transform, 213
Unit constant sequence, 250
  discrete Fourier transform, 250
Unit element:
  of set, 27
  under convolution, 226
Unit step function (stp), 217
  Fourier transform, 218

"V" function, 141
  Fourier series, 141
Vector space, 30
Venn diagram, 24

Windowing, 282
Weighting kernel, 238
Whittaker–Shannon sampling theorem, 272